Planet Earth
Problems and Prospects

In 1991, a remarkable symposium brought together humanists, historians, earth scientists, biologists, medical scientists, physicians, sociologists, and Native people to discuss the problems facing planet Earth and identify solutions and ways to avert impending catastrophes. The conference took advantage of the confluence of meetings of the Royal Society of Canada, the Learned Societies of Canada, and the Canadian Federation of Biological Societies at Queen's University in Kingston, Ontario. *Planet Earth*, a compendium of papers presented at the conference, is the result of that meeting.

JAMES A. LEITH is professor of history, Queen's University.
RAYMOND A. PRICE is professor of geological sciences, Queen's University.
JOHN H. SPENCER is professor of biochemistry, Queen's University.

Planet Earth

Problems and Prospects

EDITED BY

JAMES A. LEITH,

RAYMOND A. PRICE,

AND JOHN H. SPENCER

McGill-Queen's University Press
Montreal & Kingston • London • Buffalo

ISBN 0-7735-1292-6 (cloth)
ISBN 0-7735-1312-4 (paper)

Legal deposit second quarter 1995
Bibliothèque nationale du Québec

Printed in Canada on acid-free paper

McGill-Queen's University Press is grateful to the Canada Council for support of its publication program.

Canadian Cataloguing in Publication Data

Main entry under title:
Planet earth : problems and prospects
Papers originally presented at a conference held at Queen's University, Kingston, Ont., 1991.
Includes bibliographical references and index.
ISBN 0-7735-1292-6 (bound) –
ISBN 0-7735-1312-4 (pbk.)
1. Ecology. 2. Man – Influence on nature. 3. Economic development – Environmental aspects. 4. Environmental protection. I. Leith, James A., 1931– . II. Price, Raymond A. (Raymond Alexander), 1933– .
III. Spencer, John H. (John Hedley), 1933– .
QE1.P43 1995 333.7 C95-900277-4

Typeset in Palatino 10/12
by Caractéra production graphique, Quebec City

Contents

Figures

Tables

Preface

In 1991, as part of its sesquicentennial celebrations, Queen's University was the venue for the scheduled annual meetings of the Learned Societies of Canada and the Canadian Federation of Biological Societies. The meetings were separated by the two-day period of June 7th to 8th, which provided the opportunity to develop a symposium of interest to both organizations under the auspices of the Royal Society, the Canadian Federation of Biological Societies, and Queen's University. Dr Kanji Nakatsu of the Department of Pharmacology at Queen's University and chairman of the local organizing committee of the Canadian Federation of Biological Societies approached Dr John Spencer of the Department of Biochemistry, Queen's University, with a request to organize a meeting that would address the common interests of these varied groups. Dr Spencer asked an ad hoc group of professors at Queen's University to assist him. This group included Dr Ray Price of the Department of Geological Sciences and Dr Jim Leith of the Department of History. Other participants included Dr Roberta Hamilton of the Department of Sociology, Dr Bob Shenton of the Department of History, Mr Jack Sinnott of the International Centre, Dr Ron Lees of the Department of Community Health and Epidemiology, Dr Tom Russell of the Faculty of Education, Dr Gerry Marks of the Department of Pharmacology, representing the Canadian Federation of Biological Societies, and Dr Hugh Pross of the Department of Microbiology and Immunology, representing the Faculty of Medicine.

The title proposed for the symposium by the ad hoc group was "Planet Earth: Problems and Prospects," suggesting the general themes of global change in the geosphere-biosphere system, world hunger, world health, and the experience of the aboriginal peoples of Canada. The organization of the meetings was managed through Dr Stu Vandewater's Office of the Sesquicentennial, and an advisory committee was struck to help with the finalization of the program. This included Dr Mark Bisby of the Department of Physiology and President of the Canadian Federation of Biological Societies, Dr Dan Soberman of the Faculty of Law and the organizer of the Learned Societies meeting, Dr Alex Stewart, Department of Physics, Dr John Meisel, Peacock Professor of Political Science, and Dr Robert Uffen of the Department of Geological Sciences and Emeritus Dean of the Faculty of Engineering.

Financial support was obtained from Queen's University, the Royal Society of Canada (including the Canadian Global Change Program), and the Canadian Federation of Biological Societies. Drs Dan Soberman and Mark Bisby were very helpful in timetabling, maintaining the two-day period between the two sets of society meetings at the University, and the advisory committee helped to contact some of the speakers. The result was a very successful meeting attended by well over 350 people each day. Discussion was lively and a full transcript of the meetings and discussions has been placed in the Queen's University archives.

The editors express their thanks to the members of the ad hoc group and the advisory committee; the participants of the symposium; the Chairs of the various sessions; Dr Christopher Barnes, Vice-President, Academy III, Royal Society of Canada; Dr Anne Whyte, Director of the Social Sciences Division, International Development Research Centre and Chair of the Board of Directors, Canadian Global Change Program; Dr Kevin Keough, President of the Canadian Federation of Biological Sciences; Dr Mark Bisby, Past President, Canadian Federation of Biological Societies; and the members of Dr Stu Vandewater's Sesquicentennial Office, particularly Mrs Marcia Jones and Ms Gwen Paxton, who were responsible for assisting with advertising, contacting speakers, arranging for hotel accommodations, and for organizing use of the amphitheatre and other facilities for the symposium. Finally, thanks are due to the secretaries of the Department of Biochemistry, and to Mrs Lorraine Hellinga, who was responsible for preparing the final typescript of this volume.

JOHN H. SPENCER, Department of Biochemistry

RAYMOND A. PRICE, Department of Geological Sciences

JAMES A. LEITH, Department of History

Queen's University at Kingston, July 1992

Planet Earth

1 Humankind: The Agent and Victim of Global Change in the Geosphere-Biosphere System

DIGBY J. McLAREN

INTRODUCTION

The idea that humans, uniquely, change the face of the earth goes back almost a century and has recently been summarized by Fyfe.[1] In the 1920s, Vernadsky, a Russian geologist, began to issue warnings against what he would ultimately characterize as "a new geological phenomenon on our planet. In it for the first time man becomes a large-scale geologic force. Chemically, the face of our planet, the biosphere, is being sharply changed by man, consciously, and even more so, unconsciously."[2]

From the start there seems to have been a realization that changes should be the concern of both the natural and the social sciences, particularly in view of the early recognition that change may be both deliberate and inadvertent. The global realization that change is accelerating and may prove to have harmful results finally led to the initiation at Berne, Switzerland, in 1986, of an International Geosphere-Biosphere Program (IGBP) by the International Council of Scientific Unions (ICSU). This was followed by an international Human Dimensions of Global Change (HDGC) program. These programs coordinate, subsume, and assume a wide variety of global programs representing a veritable alphabet soup of acronyms, including many existing programs in climate, oceanography, and the use of renewable and non-renewable earth resources. The national response to global change is coordinated by the Royal Society of Canada as the Canadian Global

Change Program (CGCP), which includes both scientific and human dimensions.[3]

Evidently, we face a new problem in pursuing these many initiatives: how may what we already know be synthesized and used? Many changes are already apparent on the planetary surface, the multiple effects of which are readily demonstrated by scientific, technological, and socio-ethical interpretation. While empirical observation is a minimum requirement for policy formation, there are already ample facts from which to draw conclusions sufficient to justify, and indeed demand, immediate action. However desirable further data-collection, modelling, and interpretation may be, our first priority must be to synthesize the knowledge we already have so that policies may be formulated quickly to deal with accelerating ecological degradation.[4] The term "ecology" points to the key objective or goal. This may be summarized as the health of the planet, or, more specifically, the health of the ecosphere within which humankind and all other forms of life exist and on which they are totally dependent. The title of this chapter suggests definition of the problems that face us; possible solutions may be implied in the discussion but are not treated analytically.

I shall begin with a simple statement from the Vancouver Declaration.[5] This was made by a group of sixteen scientists and philosophers from all over the world, brought together in September 1989 at the University of British Columbia following a suggestion by UNESCO to discuss "Science and Culture for the 21st Century: Agenda for Survival." This brief declaration begins by pointing out that the discovery some two hundred years ago of fossil fuels gave us the power to dominate the planetary surface. It continues:

> In an unbelievably short span of time, unplanned and almost mindlessly, our species has become by far the largest factor for change on the planet. The consequences have been drastic and unique in the history of our species:
>
> - an accelerating increase in population growth over the past 150 years from 1 billion to over 5 billion with a current doubling time of 30–40 years
> - a comparable increase in the use of fossil fuels leading to global pollution, climate and sea-level change
> - an accelerating destruction of the habitat of life, initiating a massive and irreversible episode of mass extinction in the biosphere, the basis of the Earth's ecosystem
> - an unimaginable expenditure of resources and human ingenuity on war and preparation for war.
>
> And all licensed by a belief in inexhaustible resources of the planet encouraged by political and economic systems that emphasize short-term profit as a benefit, and disregard the real cost of production.[6]

The statement recognizes that knowledge and science could provide remedial action, and that "only the social and political will is lacking."[7]

The Declaration continues by examining the origin of our ecological problems, and gives voice to the realization that we must look for visions of a future that will allow us to survive in dignity and in harmony with our environment. It emphasises that time is short and that delay increases the cost of survival.

HISTORY

Homo sapiens has been around for about forty or fifty thousand years, and perhaps as long as one hundred thousand years. During that time, beings with a cranial capacity equivalent to our own and presumably with an intelligence similar to ours have inhabited different regions of the planet, with a population that probably never rose above a few tens of thousands in total. Food and waste, therefore, were plainly not a problem, although large (and even drastic) changes in climate and sea level controlled migrations and numbers. About three hundred generations ago (nine or ten thousand years) a slow but major change took place. This was the agricultural revolution, in which we began to plant seed and harvest crops, and to domesticate animals for food and power. As soon as we could produce surplus food, job specialization increased the rate of technological development and it was found convenient to live in one place, organizing food supply from the surrounding countryside. This led to an increase in population, although in each of the independent regions of the world where such developments were taking place very large fluctuations in population were the norm. These were due to unexpected climate change, as well as to the effects of human activity. For example, the collapse of farming systems as a result of injudicious overuse of soil and of irrigation was forced by urban growth and misguided urban leadership.[8] The whole human machine was fuelled by the sun. We ate plants and animals and used animals as well as (and almost universally) slaves for transport and agricultural work. Some hazardous wastes were produced, perhaps the most dangerous being the toxic chemicals in smoke from wood fires, and soils were poisoned through injudicious cultivation and irrigation. Solid-earth resource use was limited to building materials, ceramics, native metals such as tin and copper, and small quantities of coal and even petroleum. Technology advanced, but not the quantity of the energy supply; from the Middle Ages to the eighteenth century men and women were well clothed and could travel widely on foot, in horse-drawn carriages or on horseback, and in efficient sailing vessels. They lit their homes with animal, vegetable, or mineral oils and waxes. Wastes were still mainly self-cleansing, although cities continued to be extremely unhealthy places to live in and massive plagues swept across the world, changing demographic patterns drastically. By the eighteenth century improved agricultural practices and the beginning of improvements in hygiene increased longevity and set the scene for a more rapid increase in population. From the time of the agricultural revolution the system functioned on energy from the sun, but at a rate that was strictly limited. This allowed the population to grow to half a billion (at most) without unduly straining the resource base or the ecosphere.

The Industrial Revolution brought about the largest change in the human occupation of the planet since our race began. It is unique in history, and will remain unique until we bring about the next major revolution by coming out on the other side of our current high energy episode. It was based on a complete reversal of the existing system. We learned to use fossil fuels (coal, including brown coal and peat; oil, including heavy oils and tar sands; and natural gas). We made a Faustian bargain under which the rate of energy supply to support life was no longer restricted essentially to photosynthesis by plants but became seemingly infinite by the use of stored solar energy accumulated within the earth's crust over a period of about 400 million years. The other side of the bargain, however, was that the quantity of power stored in fossil fuels is in fact finite, and, more ominously, that the use of these fuels carries a dreadful price: the production of many poisonous materials, solid, liquid, and gaseous, of which many are new to the ecosphere, which has functioned in chemical balance for some 3.5 billion years. Matching the increase in energy use and technology during the last two hundred years was an even greater acceleration in resource use of all kinds in the Northern world. Thus began the massive discrepancy between North and South that continues today.[9] Alastair Taylor, in this volume, also takes a historical approach to the problems of today's world. We find ourselves strongly in agreement, although we approach our subject from different backgrounds and entirely independently.

I should like to examine the changes brought about by the developments I have summarized. These involve energy use and population, and the effects resulting from their continued growth. An attempt is made to characterize these problems in terms of our scientific, social, and ethical responsibilities and to consider whether common sense might play a role in guiding us in the future.

ENERGY

We are today using so much energy, largely from fossil fuels, that we can talk in terms of the global proportions of this use. Holdren[10] has given a summary of energy consumption since 1850, as well as scenarios of the future that depend on predicted population and energy use per person. Global consumption of energy is expressed in terawatt years. One terawatt year (TW year) is approximately equivalent to the energy produced by burning one billion tonnes of coal or 5 billion barrels of petroleum. By 1989, the world had used a total of nearly 400 TW years, and at the present rate of growth this will double to 800 TW years in the next twenty years. This figure is between one-third and one-quarter of the total endowment of fossil fuels. Estimates for oil and gas reserves may be found in White,[11] and for coal in Schilling.[12] Major fossil fuel reserves (in TW-yr) may be estimated as follows:

Petroleum	450
Natural gas	300
Coal	1800
Total	2550

There is little doubt that these recoverable reserves will be increased over the years, and could be doubled if future shortages force major price increases that stimulate exploration and the discovery of new reserves and also make it economically feasible to exploit lower-grade resources. The figures are, at best, merely indicators of approximate magnitude. Nevertheless, in view of the fact that the increase in use of fossil fuels to date has at least doubled every twenty years since 1890, even such gross estimation is surely a useful exercise.

Energy usage in 1990 for a population of 5.3 billion was 13.5 TW years. This represents 7.5 kilowatt-years per capita for the one billion or so people in the North, and 1.1 kW year for the more than 4 billion in the South. Industrial energy accounts for 90 per cent of these totals and derives mainly from coal, oil, and natural gas, with smaller contributions from hydro and nuclear power. The remaining 10 per cent is from traditional fuels such as wood and crop wastes.[13]

The combustion of fossil fuels of all kinds, as well as the evaporation of fuel liquids, releases poisons into the atmosphere that are transferred into the waters and lands of the continents. Obvious effects include urban air pollution, water and soil poisoning, acid rain, and the accumulation of greenhouse gases. All of these effects are linked and interdependent; all are accelerating, and are driven by increasing technological innovation, by increasing fuel use, and by increasing population. The rate of increase in energy use in the North has begun to level off,[14] but this is more than balanced by the slower and more modest increase of fossil fuel use in the South, which has four times the population. The choice of future energy supply must necessarily be influenced by CO_2 production and/or all the other deleterious effects of fossil fuel use.

If one compares figures of current use with the possible total endowment of fossil fuels one can see that the possibility of climate warming is not the only argument for switching to non-fossil sources. Regardless of pollution and climatic effects, should not chemically valuable fuels be preserved as feed-stock for future generations instead of being converted to heat in a brief blaze of glory? Most projections suggest that sooner or later we shall reduce our use of oil, gas, and coal, probably in that order, and increase non-fossil fuel use. Although there is no agreement on what the non-fossil mix will be, one can confidently claim that possibilities for alternative energy forms are probably very much better than would be suggested by the almost negligible interest and research support from government and, surprisingly, from industry. Few would disagree that there are many reasons to reduce the use of fossil fuels. Long term alternatives to these energy sources are sunlight, wind, water, ocean heat, biomass, geothermal heat, fission breeders, fusion, and advanced energy efficiency. For immediate action, massive conservation measures would appear to be the most cost effective and sensible solution pending major changes in technology.[15] Holdren[16] has given a useful discussion on the kinds of scenarios that may be drawn for a mix of varying energy levels and population densities. They all require an increase in total global energy use; this will be difficult to achieve.

POPULATION

The world population grew from about 200 million people at the time of Christ to about 500 million (half a billion) by the mid-seventeenth century. It doubled to one billion by the mid-nineteenth century, and doubled again by 1930. Since then population growth has accelerated at a staggering rate unprecedented in human history. Currently the global population stands at 5.3 billion. At present rates it will grow to 8.5 billion during the next thirty-five years. Hern[17] has characterized this growth as a planetary ecopathological process, and Paul and Anne Ehrlich[18] have made an important contribution to our understanding of the seriousness and complexity of the problem.

The speed of growth in population demands consideration of concepts of family planning and a global attack on the problem of how to achieve the emancipation of women.[19] In the past, demographers have theorized that development of improved living conditions and education would lead to a reduction in the birth rate. This may have been the case in much of Europe and North America. But Keyfitz[20] points out that population growth can prevent the development that would slow population growth. Discussions of rates of increase frequently obscure the actual figures of growth. Although rates have begun to decline, the net gain in population will continue to increase for at least another twenty or thirty years, and will grow from about 90 million a year today to 95 million by the year 2000[21]. Raven has pointed out that the population in the tropical world will grow for another two or three generations, in spite of predicted reduction in rates. Such growth in itself constitutes a threat to survival by bringing about a steady increase in poverty, the main cause of famine. Raven states: "Consequently and paradoxically, although I would subscribe to the principle that absolute numbers of people are the underlying cause of environmental destruction they are not as difficult to solve as many other aspects of the problems that are receiving very little, if any, attention."[22]

Raven is referring to the fact that if population growth ended tomorrow the tropical forests would nevertheless all be destroyed within thirty years, a quarter of the biological diversity of the world would be lost during the next fifteen to twenty years, 400 million or more people would continue to be hungry, and the sustainable productivity of the world would just as surely be destroyed.

It is important to appreciate the overall configuration of population distribution in the world. We may visualize the fundamental demographic division between North and South as a line running around the world from the Mexican–U.S. border across the Atlantic and through the Mediterranean, then over Afghanistan, Mongolia, and China, and down below Japan into the Pacific. North of the line, the population is a little over a billion; south of the line, it is a little over 4 billion. The North, however, consumes about 75 per cent of the earth's usable resources and produces the same proportion of waste – solid, liquid, and gaseous. The South uses about 25 per cent of the earth's resources and produces an equivalent amount of waste. Per capita, this works out to a ratio of 10 to 1 between North and South. These are average figures, and the disparities are very much larger in many areas and smaller in others. Nevertheless, there are few, if any, exceptions to this harsh barrier, and one should remember that the net cash flow between the two regions continues to be from South to North. The colonies are still with us.

All the problems of resource distribution and use, of increasing waste production and difficulties of disposal, and of the almost universal degradation of the ecosystem in the atmosphere, waters, and lands of the earth are due to population pressures and the imbalance of resource use between North and South. Because there are fewer people in the North per unit of land area than there are in the South, it would appear at first sight that the South was the overpopulated region of the world. In terms of resource use, however, it is evident that the population of the North should be multiplied by a factor of 10 to arrive at a per capita equivalence in regard to the depletion of nonrenewable resources, the production of wastes, and damage to the ecosphere. As we are talking about finite capacity in resource supply and most certainly in waste disposal, including aqueous and gaseous waste, we must concede that the North is overpopulated and that its current resource usage and waste production rates greatly exceed any conceivable level of sustainability. The question of the impact of population on global energy use is discussed in more depth in Holdren.[23]

There is a puzzling phenomenon currently observable relating to the nonrecognition of population growth as a fundamental danger to the planetary ecosystem. In virtually all high-level and intergovernmental discussion the question of population is ignored or receives minimum attention. It is unclear whether this seeming conspiracy of silence is due to ignorance or to design. For example, the conference goals for the United Nations Conference on Environment and Development (UNCED) and related documentation make no mention of population. Nor do the titles of any of the twenty-two related meetings that were scheduled all over the world between April 1991 and June 1992[24]. One should note that the Inter-American Parliamentary Group on Population and Development has called for fast action "essential to put Population on the Agenda of the 1992 Earth Summit in Brazil."[25] Gita Sen, in this volume, gives a wide-ranging discussion on problems of population, poverty, hunger, and sustainability.

ATMOSPHERE

There is considerable current interest in the possibility of climate warming and sea level change. Several trace gases produce a greenhouse effect in the atmosphere. These are carbon dioxide, methane, nitrous oxide, chlorofluorocarbons, and ozone. Of these, CO_2 produces about 50 per cent of the effect, but the other gases are increasing more rapidly. Measures to slow the increase and stabilize production of CFCs, also powerful greenhouse gases, have not yet taken hold. It is generally accepted that these gases are responsible for changes in the ozone layer and that these changes are therefore due to human activity: CFCs do not occur in nature. Recent information suggests that ozone depletion continues to exceed all predictions.[26] A discussion of these and related matters is given by Michael McElroy in this volume.

In 1986, analysis of ice cores from the Soviet bore hole in Antarctica at Vostok demonstrated that for the last 160,000 years the level of CO_2 in the atmosphere has correlated approximately with mean global temperature.[27] No claim is made here for a causal relationshp in either direction, but we do know that the CO_2 level in the atmosphere is now increasing steadily and is currently much higher than at any time during the last 160,000 years. The controversy over whether the warming has started or not is far from settled, but it is strange to hear arguments stressing the need for scientific certainty and better data.[28] Scientific certainty is seldom reached, but one can say in regard to climate warming that the probability that it has begun and will continue is not zero. Therefore the problem should be approached in the same manner as in normal engineering risk assessment.[29]

The automobile is primarily responsible for the high level of atmospheric pollution resulting in urban smog and greenhouse gases. It is also a remarkably inefficient means of moving people about. Measures currently suggested to reduce auto emissions will be negated if the present rate of auto production is maintained.

Acid rain (caused primarily by the high level of sulphur release from burning coal, and by nitrogen oxides from car exhaust) may soon be brought under some degree of control in North America. This may be looked on as a local or regional rather than worldwide phenomenon. Sulphur continues to increase in the global atmosphere and is responsible for widespread damage to trees throughout eastern and western Europe, where it presents a major international problem. Climate stress may also play a role in such damage but there is uncertainty as to its relative importance.

Atmospheric pollution and climate warming are inevitably linked to energy use, which is currently largely from fossil fuels in the North. These effects are also linked to social and political problems, particularly rates of use and the contrast between North and South. While governments in the United States and Canada recognize that emission reduction is necessary, they are reluctant to incur major economic costs in order to bring this about.[30]

There is a growing awareness in the South that countries attempting to increase their quota of energy may be faced with the need to adopt means of cutting back greenhouse gas emissions if energy use is increased. It is felt that much of the financial and technical burden of putting such systems in place should be borne by the countries with a high rate of fuel use and ample resources. Similar arguments are being made in regard to reduction of CFCs, where substitution will unquestionably create a financial burden.[31]

LAND

About 97 per cent of all food consumed by humans comes from the land. It is produced from 35 per cent of the total land area of the planet (4.6 billion hectares), of which one-third is in crops and two-thirds in livestock grazing.[32] The available lands are currently being mismanaged; soil degradation, which includes soil erosion and salinization from irrigation, results in a loss of some 6 million hectares each year.[33] Against this loss is the current yearly expansion of cropland of about 4 million hectares and the yearly conversion of some 10 million hectares of mostly forest land to cropland. The ecological deterioration of marginal lands is closely connected to economic growth, as well as to sociopolitical factors linked with high population growth.[34] The current global rate of soil erosion has been emphasized by Fyfe[35] and is discussed as a national environmental problem by the National Research Council of the United States.[36] Wolman[37] has shown that the sediment carried in the rivers of the world due to human activity (largely agriculture and forestry) is ten times the natural burden, and that sediments moved by farming, mining, quarrying, and construction may be ten times greater than the total quantity carried by the world's rivers to the sea.

The forests of the world covered an estimated 6.2 billion hectares at the time of the agricultural revolution, some ten thousand years ago, when clearing land for crops may have begun. Today this coverage has shrunk by one third of the total to 4.3 billion hectares. This total is diminishing rapidly as deforestation accelerates globally.[38] If present trends continue, the Brazilian rain forests, to take one example, will be cleared within the next thirty years. The importance of forests in the ecosystem can scarcely be exaggerated, given their role in photosynthesis, in acting as a CO_2 sink, and in climate modulation. In addition, the tropical forests support a huge diversity of plant and animal species in many specialized habitats.[39] For a discussion on habitat loss and biotic extinctions, see M. Brock Fenton's paper in this volume.) Finally, current forestry practices lead to an increasing generation gap between forest land harvested and land regenerated. Cleared land is subject to soil erosion and may be rendered incapable of supporting further growth. The pressures forcing deforestation are formidable. Today there are about 1.5 billion people with no feasible energy alternative who are cutting firewood faster than it can grow.

WATER

Ninety-seven percent of the earth's water is salt, and most fresh water is trapped in the ice caps. The remaining fresh water constitutes a complex system involving the atmosphere (in vapour, cloud and rain), the land surface (involving direct run-off into streams and rivers, evapotranspiration from vegetation and other land cover and lakes), and infiltration through the unsaturated zone to aquifer recharge. The hydrologic cycle interacts with terrestrial vegetation, which in turn influences the climate.[40]

Human activity has a profound effect on these interactions. Competition for water continues to increase, driven by demands for energy, transport, fishing, agriculture, irrigation, domestic use, sewage transport, and the dumping of wastes from industrial plants (including nuclear power plants, pulp and paper mills, and aluminum refineries). Dam construction, globally, has had major and largely unpredictable results on, for instance, farm irrigation and fishing.[41] Currently, ground water is being pumped from many aquifers faster than recharge, to the extent that some will be incapable of regeneration. Deforestation, particularly in the tropics, increases erosion by water 100-fold. Three quarters of the world's fresh water supply is used in irrigation.

The wetlands which occur between 50 and 70 degrees in the Northern Hemisphere constitute the second largest carbon pool (after the tropical forests) on the planet. Fourteen per cent of their area occurs in Canada. They also constitute an enormous reservoir of biodiversity, and are the largest single contributor to atmospheric methane. Human effects on the wetlands are among the most serious. Current activity is reducing wetlands rapidly and the possibility of a rising sea-level resulting from global warming may do further major damage to the coastal wetlands of the world. In summary, if global changes continue we will have to change the way we use water.

Water plays a passive role in one of the major pollution problems on earth. Old waste dumps pose a major threat to human health from the ingestion of toxic substances in drinking water. Leached from dumps, the poisons enter aquifers and spread at unpredictable rates. Waste dumps occur in unknown numbers in most areas of the world where fossil fuels are used as feed-stock for chemical industries or refineries. In addition, the new megalopolises currently in the process of formation, e.g., Mexico City, Calcutta, and New York, constitute large localized sources of pollution. Attempts to detoxify ground water by the United States Environmental Protection Agency by pumping and treatment have proved ineffective. Abelson[42] has summarized the current situation in the United States and recommended that emphasis should be given to carrying out surveys to determine the extent of ground water pollution and to creating improved technologies and better policies for the management of the problem. Of high importance, furthermore, is continuing reduction of contaminants at the primary source.

The Great Lakes constitute another major example of environmental degradation that demands regional and local concern. Attempts to deal with developing conditions have often focused on single problems such as lampreys or phosphate pollution, but these have been largely ineffective. Rather, an ecosystem approach is required.[43]

SYSTEM MALFUNCTIONS

The study of global change embraces a wide variety of disciplines and concepts. Nevertheless, the complex physical problems at issue can be classified under a few main headings:

- climate change, which is tied to energy use, affecting agricultural practices and deforestation, possible sea-level rise, changes in rainfall and terrestrial water cycle, changes in growing conditions for crops and trees, and adverse effects on health of the biosphere, including human beings
- stratospheric ozone depletion due to use of CFCs leading to stress in the biosphere and increasing cancers in humans
- acid deposition arising from atmospheric pollution from fossil fuel use, leading to biospheric and human damage
- many changes brought about by human activity involving regional and urban pollution of air, soils, surface and ground water; accelerating biota extinctions; modification of continental run-off; agricultural systems, including irrigation, that lead to soil erosion and poisoning; reduction of farmland by development.

A recent report from the International Geosphere-Biosphere Program recognizes human activity as a planetary force that drives global change. Growth in populations and in their capacity to produce goods and services is, in effect, a measure of energy use. The report then discusses the need to develop a predictive understanding of possible fates in the next decades or centuries "in order to provide decision makers with a firmer base for national and international policy formulation." Looking ahead fifty years, to a time when global change "may approach critical dimensions," they remark that there is a degree of urgency in this initiative.[44] I believe this is a misguided, if not dangerous, conception of the kinds of problems facing us. We all appear to be agreed on the nature of the physical changes that are taking place but lack agreement on the immediacy of the need for action. The curves for the forcing functions – population and energy use – are rapidly becoming steeper. Little time is left. For instance, according to the UN Population Crisis Committee, if immediate action is taken to initiate a vigorous family planning program the population might stabilize at about 9.3 billion in the next century. On the other hand, if the present rate of birth continues unchecked for the next ten years the most optimistic estimate of stabilization must be raised to 14.2 billion. Perhaps even more important is immediate reduction of growth in energy use: this must be done at a time when the South, four-fifths of the population, wishes to increase per capita energy use. In spite of major planning activities and well-designed research programs, there appears to be a lack of awareness of, or a reluctance to recognize, system-wide malfunctions and to communicate the urgency of these effectively to policy-makers.

ECONOMIC GROWTH

The concept of growth has long had currency within the North American economic system but is now being increasingly questioned. One can recognize a developing unease in some quarters in accepting mainstream economic thinking. Julian Simon remains a champion of technological opportunism, which, he claims, will overcome "Doomsday scenarios" by dint of the "Ultimate Resource," human ingenuity. Simon foresees no limits to growth in the physical sense, whether in monetary terms or in resource use. One still hears the term "technological fix" among some economists, whose touching faith is shared by few practitioners of technology.[45] An understanding that the economy exists *within the environment* grows but slowly.

Recently a new solution has been suggested to the problem of growth without limits, and the notion of growth has been redefined. By this new conception, growth in the use of raw materials and energy is superceded by an increase in value of *existing* materials by means of their rearrangement. The value of simple elements is augmented by their recombination as, for instance, in making transistors or airplanes from a relatively few chemical elements.[46] We are assured that this can be done such that physical mass is always conserved;

at the same time, the economy grows. (Does this represent a new law of thermodynamics?) There need be no limits to ingenuity in this process; it can continue indefinitely creating, as with our present system, new wants and new consumers irrespective of need. Even energy use might be reduced without impeding economic growth.

It seems difficult to reconcile this system with reality. The use of raw materials may be reduced by new technology, but recycling will remain a relatively minor factor and the law of entropy will continue to operate.[47] Energy will continue to be involved in all processes. Alienation from the biosphere by virtue of an increasing technostructure would be accelerated. Population pressures would continue to increase far faster than a new economic system could be put in place, and it is reasonable to ask what proportion of the world population might benefit by the new growth. Greater demands on renewable resources would be made if consumer living standards were raised. Nonrenewable resources and fossil fuel needs would expand for manufacturing, distribution, and use of the increasing technological advances.

As with any technology the waste products of manufacturing and disposal of used articles must be quantified and costed. Any technological innovation must, above all, be assessed for effects on the ecosystem. Given the enormous disparity of living standards between North and South, the time has come to ask of every proposed resource use and technological development: "Where?"; "For whom?"; "Why?"

THE FUTURE AND SUSTAINABLE DEVELOPMENT

Once we accept the reality of global change, we may formulate different visions of the future. One of these is typified by Ausubel's brief history of the human exploitation of the planet and of the development of an infrastructure by which we have the capacity to colonize all of it.[48] Climate change, in Ausubel's view, will not take place more quickly than we can adjust to it, and a range of technologies now available already appears to have lessened the vulnerability of human society to climate variations. In such a vision of the future there is no awareness of an alternative vision of today's moment in history, when the accelerations of the last two hundred years are approaching collapse. Weiskel also takes a historical approach but comes to a different conclusion. He may speak for the opposition: "The shrill voices of the intelligentsia in our day … seem more committed to championing their private agendas for survival than they do in extending their understanding of system-wide malfunction."[49]

A third vision of the future is typified by a middle way between technological dominance of the planet and concern for the protection of the ecosphere. This might be typified by the goals of UNCED, which involve international conventions or agreements on climate, biodiversity, and forests; an appreciation of the rights of citizens in pursuing sustainable development; and an action plan to provide for integration of environmental and economic concerns leading to new environmental laws. I believe these goals to be dangerous. They ignore the immediate crisis of population and energy use while encouraging belief in the seductive concept of sustainable development and implying that environmental concerns and the economy might be brought into balance. In effect, such an approach encourages delaying tactics.

These remarks lead to further consideration of the concept of sustainable development. This term is being used loosely to describe the goal of undefined actions that will allow humans (or some humans?) to continue to live on this planet at an undefined standard of resource use and, presumably, of waste production. Sustainability is evidently the goal of any life system, but this desirable state is surely unrealizable while the use of our planetary endowment of finite natural resources and the production of solid, liquid, and gaseous waste, including the poisons generated by the burning of fossil fuels, continue to accelerate, and while these problems are exacerbated by the runaway increase in global population and the need to reduce the huge disparity in resource use between North and South.

Munn[50] has given a useful discussion of "sustainable development," in which he emphasises the difficulty of defining this term. There is certainly a considerable difference in its use by an ecologist or an economist or an entrepreneur. Munn emphasises the danger of a stand-off between North and South. Survival strategies may be effective in areas of nearly stationary population and satisfactory living standards, but in other regions where the population is growing rapidly there might be neither strategy nor hope. Are we, therefore, to advocate regionalization in the world?

The North is free to pursue its own path of technological development and energy use and increasing production of waste. But as we try to envision our future, it would be madness not to take into account the growing disparity between North and South. It is surely reasonable to discuss this disparity, together with other interpretations of the observed facts.

Now, and in the immediate future, it might be possible to consider sustainability in a defined and limited area or region. But in the face of the disrupting forces dominating the planet as a result of population pressure and growing inequality, it would seem necessary to consider sustainability on a global scale if any useful action is to be taken. Granted, then, that if we must strive for global sustainability, we are immediately faced with the startling fact of gross overuse of resources and overproduction of wastes by the North. We must accept the necessity of attempting to include humankind as part of the total ecosphere, and to establish an equilibrium in the availability of resources if we are to continue to exist on the planet. This must be the goal of sustainability. There are, therefore, simple, unequivocal, and obvious actions that can be taken to attempt to reduce consumption and waste production in the North and to improve quality of life in the South by encouraging a modest increase in resource use while attempting to prevent further damage to the ecosphere. Plainly, it will be deeply difficult to arrive at a consensus and to achieve social acceptance of such measures. We appear to be unimpressed by immediate threats if they do not affect us personally today, even if they are certain to happen tomorrow. Sustainability must remain an objective, but it would be vain to believe that this is easily attained, or, indeed, that such a condition currently exists anywhere on earth.

There is a tendency within Northern countries to encourage belief in sustainable development, or sustainable growth, but, for reasons already given, arguments in their favour are not easily justified. In fact, with the present population of the earth at more than 5 billion, we probably lost any prospect of attaining sustainability some time ago. In 1972 Preston Cloud[51] concluded that while it was difficult to arrive at an optimum figure for the population of the planet living at a reasonable level of quality of life for all inhabitants, that figure was probably very much less than the population at the time of his writing (3.5 billion). Pimentel[52] suggests that a reasonable quality of life might be enjoyed by all on the planet if the population were about one billion.

QUESTIONING TECHNOLOGY

The problems of economic development and of the expanding use of technology are bound up with the need to question existing and new technologies.[53] We must go back to the beginning of the Industrial Revolution to understand the powerful thrust given to technological development by the discovery and widespread use of fossil fuels. Today we find ourselves faced with the need to question the growth of new technologies and contemplate the growing harm the burning of fuels inflicts on the global environment.

We cannot question the conclusion that "something has gone wrong," nor can we doubt that our present predicament has been brought about, essentially, by the unrestrained development of technology supported by the seemingly limitless supply of energy generated by burning fossil fuels. We are thus faced with a fundamental moral or ethical dilemma. If we concede that many of the technologies developed during the last two hundred years have been responsible for major damage to the ecosphere, then scientists, engineers, and technologists must with the wisdom of hindsight ask themselves whether they have been negligent in their duty to warn society against the misuse or misapplication of their discoveries. Are scientists and technologists responsible for the use to which their discoveries and technologies are put? This question is not easily answered. But if scientists assume no responsibility to whom must that responsibility be assigned? Not to policy-makers or politicians, for they could have had no knowledge of the contents of Pandora's box before it was opened. Probably not to the public, although they may have provided the incentive for new gadgets and means of gratification to be devised. Perhaps, in some instances, we may lay the blame on the shoulders of those who funded and encouraged scientists and technologists in their work and sought power or profit in return. But we can scarcely claim that this absolves the scientific community from its share of the responsibility.

By like argument, are scientists and technologists ready to ask themselves whether they should take responsibility for explaining the results of current researches and inventions before they generate further wants and more dangerous products? If, indeed, technology is misapplied, then all existing and new technologies must be evaluated with reference to damage to the ecosphere, sustainability of resource supplies, and waste disposal.

It appears that scientists and purveyors of knowledge of all kinds should be worried about the widespread misapplication of their discoveries. They should be the responsible trustees of new knowledge and technologies. Public decision-makers also fail to take intelligent account of available knowledge which bears on policy, and thus contribute to its misapplication. Furthermore, there are many difficulties in obtaining objective or disinterested opinions from individual scientists within disciplines. The big questions are avoided, and disciplinary loyalty is expected within each academic department. The idea of productive synthesis is not well developed. But any branch of science responsible for producing new data or interpretations has an obligation to communicate the importance of these to the public. If this is not done, then one must question whether the research is justified. Transdisciplinary knowledge synthesis – and this must include the social sciences – is required to formulate policy to deal with accelerating ecological degradation. Examples from the past reinforce our realization of the dangers of making assumptions in areas of public policy without an adequate synthesis of scientific, social and moral aspects.[54] Environmental challenges induced by increasing technological developments are of global significance. Yet the world seems far from generating political structures capable of coming to terms with the gravity and pervasiveness of environmental decay.[55]

Political, social, and even scientific debate must deal increasingly with the problem of norms. On what norms should a pluralist society be based? In the light of ongoing changes in technology can society produce appropriate ethical standards that can be borne out in the economic and political sphere? For instance, we should not ask "Is science effective?" but rather "What kind of science should we have? Is this science (or research) necessary to a particular sphere of social behaviour?" Many perspectives are not being addressed in attempts to arrive at syntheses that include science, social sciences and ethics. Perhaps some system involving setting up a "Science Court" could be envisaged that would help individuals to deal with, and indeed recognize, the ethical problems that can arise as they serve the interests of their employers.[56]

All our ethical considerations may be reduced to two general principles. First, a code of ethics that is seen to be humane and just in a human framework may result in increased stress on the ecosphere. Second, only human actions that also benefit the ecosphere (of which we are a part) may be considered ethically acceptable.[57] The most urgent of these actions must be a reduction in the population of human beings and in their use of fossil fuels.

There are many ethical issues that interact with the technical and social problems under discussion. Many are bound up with disparity in resource use, poverty and starvation, the plight of children, and the importance of "investing in women" in a new world order. The last word should be given to the South. Accordingly, I take my final remark from Carlos Gutierrez's discussion on the question of an adequate global model to meet the demands of survival for all human beings in a world shared in common: "The industrial nations must stop viewing the systemic imperatives of their own expansion as imperatives of nature and admit that the modification of production that is indispensable if the biosphere is to be saved requires not only an end to lineal quantitative growth but a radical reduction of economic privileges as well."[58]

NOTES

1 Fyfe, W.S. 1990a. The International Geosphere-Biosphere Programme and global change: an anthropocentric or an ecocentric future? A personal view. *Episodes* 13(2): 100–2.

2 Vernadsky, V. 1945. The biosphere and the noösphere. *American Scientist* 33(1).

3 Further information may be obtained from *Delta*, the newsletter of the CGCP.

4 Weiskel, T.C. 1991. Common sense in uncommon times: the perversion of science in periods of social stress. Paper presented to American Association for the Advancement of Science, 18 Feb. 1991, Washington, D.C.

5 Vancouver Declaration 1989. *Science and Culture for the 21st Century: Agenda for Survival*. Canadian Commission for UNESCO. (Declaration distributed 1989, volume published 1990.)

6 Vancouver Declaration 1989, 11–12.

7 Vancouver Declaration 1989, 12.

8 Weiskel 1991.

9 Rifkin, J. 1991. *Biosphere Politics*. New York: Crown.

10 Holdren, J.P. 1990. Energy in transition. *Scientific American* (Sept.): 157–63.

11 White, D.A. 1987. Conventional oil and gas resources. In *Resources and World Development*, ed. D.J. McLaren and B.J. Skinner. John Wiley & Son, Dahlem Konferenzen, 113–28.

12 Schilling, H.-D., Wiegand, D. 1987. Coal resources. In *Resources and World Development*, ed. D.J. McLaren and B.J. Skinner. John Wiley & Son, Dahlem Konferenzen, 129–56.

13 Holdren 1990.

14 Fulkerson, W., Reister, D.B., Perry, A.M., Crane, A.T., Kash, D.E., Auerbach, S.I. 1989. Global warming: an energy technology R&D challenge. *Science* 246: 868–9.

15 Fickett, A.P., Gellings, C.W., Lovins, A.B. 1990. Efficient use of electricity. *Scientific American* (Sept.): 65–74; Harvey, L.D.D. 1990. *Carbon Dioxide Emission Reduction Potential in the Industrial Sector*. Ontario Select Committee on Energy. Ottawa: The Royal Society of Canada; Abelson, P.H. 1991. National energy strategy. *Science* 251: 1405.

16 Holdren, J.P. 1991. Population and the energy problem. In *Population and Environment: a Journal of Inderdisciplinary Studies* 12(3): 231–55.

17 Hern, W.M. 1990. Why are there so many of us? Description and diagnosis of a planetary ecopathological process. In *Population and Environment: a Journal of Interdisciplinary Studies* 12(1).

18 Ehrlich, P.R., Ehrlich, A.H. 1990. *The Population Explosion*. New York: Simon & Schuster.
19 Sadik, N. 1989. Investing in women: the focus of the nineties. In *The State of World Population 1989*. New York: United Nations Population Fund.
20 Keyfitz, N. 1991. Population growth can prevent the development that would slow population growth. In *Preserving the Global Environment*, ed. J.T. Mathews. New York and London: Norton, 39–77.
21 Demeny, personal communication.
22 Raven, P.H. 1989. Managing tropical resources: a challenge to us all. In *A Modern Approach to the Protection of the Environment*, ed. G.B. Marini-Bettolo 130–55, and discussion 531–2. Pontificiae Academiae Scientiarum, Scripta Varia 75.
23 Holdren 1991.
24 Canadian National Secretariat UNCED '92. Newsletter, 1991. 1(1).
25 Inter-American Parliamentary Group on Population and Development 1991. Fast action essential to put population on agenda of 1992 "Earth Summit" in Brazil. Bulletin 8(4). New York.
26 Kerr, R.A. 1991. Ozone destruction worsens. *Science* 252: 204.
27 Crowley, T.J., North, G.R. 1986. *Paleoclimatology*. New York: Oxford University Press.
28 For example, in Brookes, W.T. 1989. The global warming panic. *Forbes* (Dec. 25): 96–101.
29 This is discussed in Bruce, J.P. 1991. Myths and realities of global climate change. *Ecodecision, Environment and Policy Magazine* (Montreal) 1(1): 89–92.
30 Abelson 1991; Canadian Council of Ministers of the Environment 1990. *National Action Strategy on Global Warming*. CCME.
31 Agarwal, A., Narain, S. 1991. Global warming in an unequal world. In *Centre for Science & Environment*. New Delhi, India, 1–36.
32 Pimentel, D. 1987. Technology and natural resources. In *Resources and World Development*, ed. D.J. McLaren and B.J. Skinner. John Wiley & Son, Dahlem Konferenzen, 679–95.
33 United Nations Environment Programme 1980. *Annual Review*. Nairobi, Kenya: UNEP.
34 Ayyad, M.A. 1989. Conservation of marginal lands. In *A Modern Approach to the Protection of the Environment*, ed. G.G. Marini-Bettolo. Pontificiae Academiae Scientiarum, Scripta Varia 75, 215–41.
35 Fyfe, W.S. 1990b. Soil and global change. *Episodes* 12(3): 249–53.
36 National Research Council (U.S.) 1989. Alternative agriculture. In *Board on Agriculture*. Washington: National Academy Press.
37 Wolman, M.G. 1990. The impact of man. *Eos* (Dec. 25).

38 Maini, J.S. 1990. Forests: barometers of environment and economy. In *Planet Under Stress*, ed. C. Mungall and D.J. McLaren, Toronto: Oxford University Press, 168–87.
39 Myers, N. 1984. *The Primary Source: Tropical Forests and Our Future.* New York: Norton.
40 Sellers, P.J., et al. 1990. Water–energy–vegetation interactions. In *Research Strategies for the U.S. Global Change Research Program* (Chairman: H. Mooney) Washington: National Academy Press, 131–63.
41 Schindler, D.W., Bayley, S.E. 1990. Fresh waters in cycle. In *Planet Under Stress*, ed. C. Mungall and D.J. McLaren, Toronto: Oxford University Press, 149–67.
42 Abelson, P.H. 1990. Inefficient remediation of ground-water pollution. *Science* 250: 733.
43 Holling, C.S., Bocking, S. 1990. Surprise and opportunity: in evolution, in ecosystems, in society. In *Planet Under Stress*, ed. C. Mungall and D.J. McLaren. Toronto: Oxford University Press, 285–300. A recent example of the ecosystem approach is Crombie, D. (Commissioner) 1990. *Watershed*. Second Interim Report. (August). Royal Commission on the Future of the Toronto Waterfront.
44 International Geosphere-Biosphere Program 1991. Global change system for analysis, research and training (START), ed. J.A Eddy, T.F. Malone, J.J. McCarthy and T. Rosswall. *Global Change Report* 15, 2, 3. Boulder, Colo.: IGBP.
45 Wilde, K.D. personal communication, 1990.
46 Romer, P.M. 1990. Endogenous technological change. *Journal of Political Economy* 98(5): S71–S101.
47 Georgescu-Roegen, N. 1971. *The Entropy Law and the Economic Process.* Cambridge, Mass.: Harvard University Press.
48 Ausubel, J.H. 1991. Does climate still matter? [commentary] *Nature* 350 (25 April): 649–52.
49 Weiskel 1991.
50 Munn, R.E. 1989. Sustainable development: a Canadian perspective. Submitted to *The Queen's Quarterly*, September.
51 Cloud, P.E. 1972. Resources, population and quality of life. In *Is There an Optimum Level of Population*, ed. S.F. Singer. New York: McGraw-Hill.
52 Pimentel 1987.
53 Wilde, KD. 1991. Economists and engineers: can we control this runaway team before they wreck our wagon? *Conference on Preparing for a Sustainable Society.* (Submitted to conference proceedings.) Toronto: Ryerson Polytechnical Institute.
54 Weiskel 1991.

55 Falk, R. 1989. *The Capacity of International Law to Respond to the Environmental Challenge.* Proceedings, 18th Annual Conference. Ottawa: Canadian Council on International Law, 35–49.
56 Wilde, personal communication 1990.
57 Rowe, J.S., Professor Emeritus, University of Saskatchewan.
58 Gutierrez, C.B. 1990. Ethics, politics and economics on an Amazon safari. In *Ethics and Environmental Policies*, 1st International Conference, Brazil, August 1990.

2 Our Common Future: World Development and the Environment

ALASTAIR M. TAYLOR
DUNCAN M. TAYLOR

INTRODUCTION

Eighteen forty-one was a momentous year for Kingston. The city had been chosen the seat of the new Parliament of Canada, and although this honour was withdrawn three years later, most fortunately the founding of Queen's proved to be permanent. The University began in one house, with two professors and ten students. A century and a half later, it has eighty-six buildings on two campuses, eight faculties and schools, a staff of 3,800, and a total enrolment of 17,668 students. During those same 150 years, Upper Canada, with 480,000 inhabitants, has become Ontario, with a population of 8.84 million and a gross domestic product (GDP) of over $273 billion (1989).

In keeping with the expansionist world-view that prevails in our society today, such growth has been equated with progress. But is "progress" the right term with which to define growth? Worldwide, population and industrialization have both increased exponentially – but so has global pollution, while the biosphere's resources required to sustain this demographic and economic growth are decreasing at an alarming rate. Since, in Dr McLaren's succinct phrase, humankind is at once the "agent and victim of global change," we have now to address a critical question: Can we sustain indefinite development – especially economic development – at current rates of societal growth and resource usage, and do so within an environment that must itself remain sustainable if humankind is to survive?

HOW DID WE GET INTO OUR PRESENT PREDICAMENT?

As pictures from outer space attest, our planet is indeed beautiful – but finite. Its geospheric and biospheric systems evolved within a vast time-scale. When the hominids appeared, the geosphere and biosphere had already been "in place" for billions of years, and were systematically functioning by means of homeostatic mechanisms. *Homo* had been provided with an endowment comprising inorganic and organic capital on a global scale. What has our species done with that capital?

Major shifts in global societal evolution have been marked by factors responsible for quantum shifts to new levels of organization. We might identify four of these to show how they have combined during the 150-year period of Queen's existence to result in a new quantum shift of planetary proportions.

The first factor comprises major scientific discoveries or technological advances. As Alfred North Whitehead remarked in his *Science and the Modern World* (1925), "The greatest invention of the nineteenth century was the invention of the method of invention." Our own century has been marked by what Norbert Wiener, the "father of automation," called the "Second Industrial Revolution" of cybernetics, information-gathering, and communication. This revolution has spawned new industries – aeronautics, electronics, computers, plastics – all of which rely on research and development for their continued advancement. Surely more scientists and technologists are living today than have lived in all previous centuries combined.

The second factor is our resulting capability to control the environment. Hitherto, humankind lived on "flat earth." But, with the Wright brothers, we made a revolutionary leap into the third dimension, initiating our ascent into the atmosphere and beyond the outer space. Concurrently, descent into the hydrosphere or inner space of ocean beds and continental shelves became possible. This expanded control capability now cuts across all physical and societal boundaries in the global environment, carrying with it irreversible consequences for the nation-state system. In effect, contemporary technological innovations are progressively shifting societies everywhere, despite their *de jure* status of national independence, to one of *de facto* transnational interdependence.

A third factor responsible for this societal revolution is an exponential increase in energy production and consumption. Studies indicate that daily per capita energy consumption among Stone Age food-gatherers was some 2,000 kilocalories (kcal). The rate increased to 12,000 kcal among farmers in 5000 BC, while by AD 1400 energy consumption may have doubled again among advanced agricultural societies. The Industrial Revolution enabled per capita consumption to rise by 1870 to 70,000 kcal. A century later, the average North American was using 230,000 kcal every day at his place of work, at home, and in his car.[1]

Accompanying these factors is a spectacular contemporary manifestation of growth: global population. Our species numbered no more than one billion when Queen's was founded. Global population has now surpassed 5.5 billion and could nearly double within the next sixty years. The United Nations Fund for Population Activities warns that projections for reaching a stable population of 10.2 billion within one hundred years "may have been too optimistic." The 10 billion mark could be reached by the year 2050, while growth will continue for another century unless substantial progress is made in reducing fertility. Meanwhile, the world's population is growing at a rate in excess of 90 million annually (which is more than three times Canada's population).[2]

Our Stone Age ancestors had such simple tools that for scores of millennia they did little more than live off the interest of global environmental capital. From a systems standpoint, just as the evolution of the geosphere and biosphere relied on homeostatic processes, so lithic societies had behaviour patterns which maintained an overall balance with their environment. In contrast, our economic and technological systems are driven by positive feedback processes – hence the "growth" dynamic – and have progressively diminished the planet's inorganic and organic capital. This expansionist ethos regards material growth as indispensable to our pursuit of happiness. Its absence, we are warned, can only result in loss of incentive, stagnation, and mass unemployment. The paradigm of continuous growth – measured quantitatively by a society's gross national product (GNP) – is based upon a critical assumption: that there will always be an unfailing supply of resources. This assumption is rejected by ecologists as the "myth of superabundance."

Our accelerating misuse of the global environment is compounded by a deep societal split between the rich industrial North and the poverty-stricken South. Customarily, societies have interacted with and enriched one another. For example, the ancient riverine civilizations, located in what is now the South, contributed a rich legacy of inventions that over the centuries were brought to Europe. The Europeans, utilizing four key Chinese inventions – paper, printing, gunpowder, and the compass – embarked with Columbus, da Gama, and their successors upon oceanic exploration and political expansion.

From the resulting colonial empires came a cornucopia of resources which, with the advent of the Industrial Revolution, enabled the "developed" countries of the North to establish a global economy and to control the South's pre-industrial societies.

In this winner–loser relationship, the North's prosperous peoples have thought that they could insulate themselves from the South's poverty and related problems. But this complacency is false: developments in recent decades are making the two sectors of global society mutually vulnerable, especially in the following four areas:

1 The environment. We need hardly belabour the obvious: problems of environmental degradation, including climate change, are global in scope. Our environmental predicament will be worsened still further should China and India – whose combined populations in the year 2000 will be double that of the North – employ current technologies to make a quantum increase in energy production as part of their industrial strategy to achieve unprecedented growth for their peoples.
2 Economics. The economic gap between North and South has become huge. Some 70 per cent of world income is produced and consumed by 15 per cent of the word's population. Today, living standards in Latin America are lower than in the 1970s, while in Africa they have fallen to the 1960s level. In their efforts to modernize and develop their economies, countries in the South have borrowed well in excess of a trillion dollars – yet in 1988 they transferred to the North U.S. $43 billion more than the development assistance they received, a situation described by Ivan Head as "a perverse, unsustainable transfer of wealth from poor to rich."[3]
3 Demographics. The rich North gets richer, the poor South gets children. And lots of them, so that in the next century five-sixths of the world's population will be in developing countries. In the year 2000, the South will have 45 of the 60 largest cities, and 18 of these 45 will have more than 10 million inhabitants each. Thirty-five per cent of the South's population will be under age 14. As the South's population grows, the North will be confronted by increasing pressures to relax immigration restrictions, thereby risking new domestic tensions of its own.
4 Politics. All of these problems exacerbate political instability in the South. But this instability can in turn involve the North, as proved by the Gulf War, in which the permanent members of the United Nations Security Council were the chief arms suppliers. For millenia, nations have armed themselves in their attempts to be secure. But "security" has taken on a new dimension. No longer can any

nation-state immunize itself against social and environmental problems by relying primarily on traditional military methods. This is because these problems are embedded in a global ecology and human community and, since they are transnational, will penetrate all political boundaries. Without equitable access to food, health, education, and economic opportunity, no society can feel secure, as the violence and unrest that arose from the former apartheid regime in South Africa so dramatically attests. This lack of equity is largely responsible for the South's instability. Unless corrected, social turbulence could become progressively immune to any traditional and simplistic reliance on armed force, as witnessed in the 1990s in states such as Rwanda, Somalia, and Haiti. For South and North alike, the bell tolls.

IS "SUSTAINABLE DEVELOPMENT" SUSTAINABLE?

Our physical growth has already done extensive damage to the geosphere-biosphere system. Have we today the knowledge, and the foresight, to reconcile *sustained* with *sustainable* development? This brings us squarely to the thesis propounded by the World Commission on Environment and Development in *Our Common Future*, the highly influential "Brundtland Report." The Commission was mandated by the UN General Assembly to "propose long-term environmental strategies for achieving sustainable development by the year 2000 and beyond."[4] Our critique addresses a dual question: was it realistic to formulate strategies conceived within the traditional expansionist paradigm, and, if so, can they be implemented?

The most positive contribution of *Our Common Future* was to marry ecology and economy on a global scale, and to focus on the critical problems resulting from that relationship. With irrefutable logic it links the environment–development nexus to the crisis of endemic poverty in the South. On the other hand, it fails to make concrete what is expected from the geosphere's and biosphere's resource endowment to meet the demands of increased economic activity for the South, as well as of continued growth in the North. Again, the Report relies on the concept of managing resources to obtain "maximum sustainable yield." But, as one critic points out, apart from the difficulties of quantifying "maximum" sustainability, past experience shows that such yields "are so bent by wishful thinking, corporate pressures and ignorance that they are usually many times too high. The example of Atlantic Canada's fish stock estimates and catch quotas is tangible proof of this point."[5]

At the heart of the Report is the thesis that global society must achieve a "sustainable development" that "meets the needs of the present without compromising the ability of future generations to meet their own needs."[6] "Sustainable development" has become the fashionable slogan adopted by a range of groups espousing conflicting interests and values – from environmental preservationists and deep ecologists to advocates of resource development and neoconservatives.

We challenge the fundamental premise on which this term is based, namely, that a development strategy that retains the expansionist world-view which paid such munificent dividends in an earlier stage of the North's development is still possible. We contend that "sustainable development" is an oxymoron as long as it is interpreted in the light of our dominant world-view and current economic assumptions. Put simply, our objection is as follows: although our economies are predicated upon the need for continual growth, the ecosystems in which they are embedded are not. Consequently, the "consumption of ecological resources everywhere has begun to exceed sustainable rates of biological production."[7] One highly visible example is found in the overharvesting of British Columbia's forests.

To repeat a metaphor from economics: we are living off the planet's environmental capital when we should be living off its interest. This is illustrated by the factor of net primary production (NPP). It is estimated that nearly 40 per cent of terrestrial NPP (from photosynthesis) has already been appropriated by humans. Given the fact that *Homo sapiens* is but one of millions of species, a doubling of the human population together with a further massive increase in economic activity could leave no more NPP for these other species, without which humans themselves cannot survive.[8] But while ecologists warn that we cannot afford to keep doubling our appropriation of NPP, the Brundtland Report adopts a different tack: "Given expected population growth, a five- to tenfold increase in world industrial output can be anticipated by the time world population stabilizes sometime in the next century."[9] What does this imply – say, in the field of energy, the future development of which must increasingly depend on "sources that are dependable, safe, and environmentally sound,"[10] even though no such mix of sources is presently at hand? Employing as an energy unit the terawatt year (TW year), the amount of energy that would be released by burning one billion tonnes of coal, the Report estimates that in 1980 global energy consumption amounted to some 10 TW years. If per capita use were to remain at the same level, by 2025 a global population of 8.2 billion would need 14 TW years (over 4 in developing and over 9 in industrial countries), an increase of 40 per cent over 1980. "But if energy consumption per head became uniform world-wide at current industrial country levels, by 2025 that same global population would require about 55 TW [per year]."[11]

What might we deduce from these figures? First, to keep this global energy consumption increase to 40 per cent, the present disparity in North–South economic standards must be maintained. Secondly, if the South insists on attaining economic parity with the North, global energy consumption must increase 5½ times. Thirdly, as of 2025, global population will still be decades away from stabilization. Meanwhile, the Report recognizes that energy in the present era "has been used in an unsustainable manner," and admits that "[a] generally acceptable pathway to a safe and sustainable energy future has not yet been found."[12]

What else does this anticipated five- to tenfold increase in global economic activity suggest? When the Report appeared in 1987, this activity had reached $13 trillion, with most of it controlled by about one-fifth of the world's population. Just to enable the South to reach the North's present resource usage and GNP would call for a fivefold increase, a global economic activity of some $65 trillion. But if, meanwhile, the North doubles its own growth (a very modest increase, considering that industrial production "has grown more than fifty-fold over the past century," with four-fifths of this growth since 1950),[13] we arrive at the tenfold increase hypothesized in the Report, bringing global economic activity to, say, $130 trillion. Either figure must dramatically exacerbate the present environmental crisis – five-fold? tenfold? – yet is still deemed by the Report to be within the planet's carrying capacity. Is this exponential growth expected to continue into future centuries? Sufficient unto the day be the myth of superabundance?

Some years ago, we encountered the term "demographic inertia" in reference to the momentum of the current population explosion. Demographic inertia is analogous to a supertanker moving with a full cargo. Even with its engines shut down, the tanker must still travel miles before coming to a halt, unless its engines are reversed to hasten stoppage. What will it take to overcome the thrust of "societal inertia" in order to achieve sustainable development? Will the engines of economic and demographic growth have to be reversed somehow to save the environment? At the first United Nations Conference on the Human Environment, held in 1972 (the Stockholm Conference), countries from the South rejected the limits-to-growth thesis as a highly sophisticated form of neocolonialism. They were prepared to accept environmental pollution – the rich countries' disease – in order to industrialize and to improve their living standards. In calling for a markedly substantial increase in the South's development, the Brundtland Report never raised a related question: whether the North may have to cut back on its own rate of expansion to make this growth possible. So far, the industrial economies have not considered even the possibility of any such cutback.

CANADA: AN EXAMPLE OF THE NORTH'S GROWTH SYNDROME

From the days of the first European colonists, the society that came to be called Canada embraced an expansionist world-view in which "progress" was equated with increased production. With little regard to the aboriginal peoples already present, this society came to occupy a vast and resource-rich land mass and enjoyed remarkable physical, political, and economic growth. Canada was a nation founded on the myth of superabundance. But this comfortable and uncritically acquired belief is no longer tenable.

In its ratio of population to resources, Canada has been blessed as very few countries have been. Yet its natural endowment *is* limited. We grew by adopting the role of hewers of wood and drawers of water. But alas, we have hewed forests in amounts and ways which are now unsustainable. British Columbia produces almost half of Canada's annual volume of timber and pulp. In 1988–89, the province cut 89.05 million cubic metres, far above the Ministry of Forests' figure of 59 million cubic metres of long-term sustained yield.[14] And, increasingly, the American sunbelt – and especially drought-stricken California – pressures us to draw on its behalf what could one day become unsustainable amounts of water. We have been told that water withdrawals in Canada are projected to rise from an estimated 120,000 million litres per day (mLd) in 1980 to 282,000 mLd by the year 2000.[15] Already we have six water-deficient basins, including Southern Ontario.[16] By itself water does not flow on a "level playing field," but our southern neighbours seem more than willing to assist its bulk movement across the border.

Canadians are concerned about their deteriorating environment. But that concern seems to wane in times of widespread employment. How do we reconcile ecological sustainability with sustained economic growth? Ottawa's response to the Brundtland Report's call for a sustainable development strategy has been subsumed within the expansionist world-view. Little wonder that Canadian business leaders support the report of the National Task Force on Environment and Economy, which "does not require the preservation of the current stock of natural resources or any mix of human, physical and natural assets. Nor does it place artificial limits on economic growth, provided that such growth is both economically and environmentally sustainable."[17]

But what if "such growth" is not "both economically and environmentally sustainable"? Will Canadian business then reverse its traditional priorities and behaviour from sustaining the economy to sustaining the environment? As one resource ecologist points out: "The Task Force is reluctant to admit the possibility that living standards for some may have to be reduced that others may live at all. It avoids this issue entirely." Nor does it recognize that preserving certain "mixes" of ecological resource systems "may well be essential to sustainability."[18] Here we see the danger that the term "sustainable development," already being coopted by established interests, will be used to call for bandage solutions which mask the need for basic political and economic changes – changes essential to maintain environmental integrity while achieving greater societal equity in South and North alike.

As Canada's environmental control expanded, *terra incognita* was transformed into new areas of settlement, while the discovery and exploitation of the riches of the geosphere and biosphere grew at a rate which more than kept pace with demographic increase. We might examine the growth of two phenomena: the federal debt, and the GNP/GDP.

The federal debt

After Canada's first seventy-three years, the federal debt in 1940 amounted to $3.27 billion; twenty years later, it had almost quadrupled to $12.08 billion. But the 1980s were the decade of spectacular increase in which the deficit soared to $358 billion. In 1980, the debt comprised 26 per cent of the GNP; in 1991 it had risen to 59.8 per cent, and there was every prospect that by 1995 the federal debt would amount to three quarters of all wealth produced in the country. The government's federal budget for 1991 assumed that by 1995–96 the annual deficit would be reduced from the existing size of more than $30 billion to only $6.5 billion. In stark contrast, however, the February 1994 budget deficit climbed to $45.7 billion, and the succeeding government's new strategy is now to try to cut the deficit in the next three years to $25 billion. Difficult as meeting this goal will be, because it means cutting deeply into unemployment insurance payments, federal assistance to social welfare programs, and post-secondary education, the deficit will then have to be cut by a further $8 billion a year in order to be eradicated by the turn of the century.

Even if this Herculean labour can be accomplished, Canadians will still be left with having to grapple with a federal debt amounting to $635.4 billion. This is one and one-quarter times larger than a previously analyzed debt (based on 1991 figures), the servicing of which over 50 years called for Canadians to pay some $2.5 trillion.[19]

Let us again be optimistic and assume that by then we shall have zero annual deficits. We must still account for the principal and compound interest on $500 billion. We have two nondefaulting options. First, we can continue simply to service the debt. Assuming that the population plateaus at around 30 million in the next century, and that interest rates are on average about 10 per cent, over fifty years we and our successors must pay $2.5 trillion, or $83,332 for every Canadian. Second, Ottawa could opt to amortize the debt (a much more sensible choice), in which case the cost of paying off the debt would amount to $2.521 trillion over a fifty-year period.

Gargantuan though this debt load is certain to be for 30 million (or fewer) Canadians, it will be accompanied by further major problems.

1 If current trends continue, the federal debt could rise to, say, 70 per cent of GDP, lowering the living standards of future generations.
2 The country's population will start to decline by the year 2025. And it will be older, so that the proportion of working-age earners will fall. Can a diminishing work force produce enough to meet the needs of a growing senior citizenry as well as satisfy its own aspirations? (And will it be able to meet the challenge of multinational corporations' high-tech productivity utilizing the hungrier work forces of the South, including that of the south bank of the Rio Grande?
3 We have also to factor into the GDP–debt equation the costs of exploiting the environment. By one calculation, just to enforce existing governmental standards over the next ten years could cost $70.2 billion. Who will pay: polluters? public? governments? Conversely, what will be the socioeconomic costs of *not* cleaning up the environment? And who will foot the bill for overhauling urban environments and infrastructures, estimated at some $15 billion?
4 So far we have concentrated on the federal debt. But the provincial deficit/debt situation is as critical. In 1993, Saskatchewan's debt had risen to $15 billion and British Columbia's to $23 billion, while Ontario's debt had soared from $30.4 billion in 1984–85 to $78.6 billion some ten years later – an increase of 254 per cent. In 1988, the combined federal and provincial public debt was $442 billion, or 72 per

cent of GDP. In 1993, it has risen to $696 billion, or 97 per cent of GDP.[20] Should this rate of growth continue, the combined federal and provincial public debt will approach $1 trillion by the end of the century – and be in the neighbourhood of 125 per cent of GDP. Already, "Canada has joined the Third World in terms of the magnitude and severity of its total all-government debt problem. All the conditions for a financial crisis of major proportions are now in place."[21]

The GNP/GDP

Confronted by these massive fiscal problems, governments remain committed to traditional economic policy. That is, they continue to seek solutions along the road of sustained growth. But just what kind of growth rate will be required? We should consider the historical record. Over a sixty-two-year period, Canada's GDP increased (in current dollars) 128 times between 1926–27 and 1988–89.[22] Between 1970–71 and 1986–87, the average annual growth was 11.4 per cent. Yet this was the same period when Ottawa ran an unbroken series of annual deficits, raising the federal debt from $18,356 billion to $264,101 billion, and the debt's percentage of GDP grew more than 2½ times. Given this track record, how can Ottawa hope to stabilize, far less reduce, the public debt? It is utterly unrealistic to extrapolate our future economic growth at the rate of the past six decades. (This would mean having a GDP of some $84 trillion around the year 2050.)

Two facts are painfully clear:
- We are living beyond our economic means.
- We are living beyond our environmental means.

In both respects, we are consuming "capital" in ever greater amounts rather than living off the "interest" from that capital. Our economic capital, created by human agency, and our environmental capital, created by nature's evolutionary processes, should comprise the endowment of future generations of Canadians. The combined misuse of our economic and environmental wealth must prove devastating unless radical remedial measures are undertaken in both areas, jointly, in the immediate future.

WANTED: A STRATEGY OF REMEDIAL ACTION

We can best begin with a realistic definition of "sustainable development." Otherwise, this fashionable term will continue to be, in the metaphor used by Maurice Strong, former head of the CIDA and member of the World Commission on Environment and Development the fig-leaf behind which business hides. As we have seen, it has been interpreted in terms of maintaining and prolonging growth, primarily in material and economic terms. But instead of sustaining development *per se*, we need to develop sustainability in the sense of that which "supplies with necessities or nourishment" – in other words, the myriad natural systems and life-support processes of this planet. Hence, human activities must not jeopardize or exceed the carrying capacity of those natural systems in which they are embedded.[23] From this emphasis on environmental sustainability emerge the following principles of sustainable development:

1 *Limit human impact on the living world to a level within its carrying capacity.* Human impact is the product of the number of people multiplied by how much energy and raw materials each person uses or wastes. An ecosystem's carrying capacity is its capacity to renew itself or absorb wastes, whichever is less.
2 *Maintain the stock of biological wealth:* e.g., by (a) conserving ecological processes that sustain the productivity, adaptability, and capacity for renewal of lands, waters, air, and all life on earth; (b) conserving biodiversity; (c) ensuring that all uses of renewable resources are sustainable; and (d) minimizing the depletion of non-renewable resources.
3 *Promote long-term economic development that increases the benefits from a given stock of resources and maintains natural wealth.*
4 *Aim for an equitable distribution of the benefits and costs of resource use and environment management* – between rich and poor countries, and between present and future generations.
5 *Promote an ethic of sustainability* – conversely, values that encourage unsustainable practices need to be changed.[24]

These principles have significant implications. One relates to restructuring the North–South relationship. As the Brundtland Report states, many of the North's development paths are not sustainable, while we in turn contend that the North, where by far the largest amount of the fortyfold increase in industrial production since 1950 has occurred, must not only lessen its monopoly on global resources but will also have to cut back drastically so that the South can survive. It is intolerable that 26 per cent of the world's population should continue indefinitely to consume 80–86 per cent of the planet's non-renewable resources and 35–53 per cent of global food products.[25]

How would such a cutback affect Canada, in view of the fact that its economy's "health" has been predicated upon continual growth? Much of this growth derives from unsustainable resource exploitation and environmental degradation. Canadians use more energy and produce more solid waste per capita than any other people. Is it not time to acknowledge that Canada's real wealth in terms of natural resources is being degraded for the sake of rising incomes and GNP, and to realize also that the downstream environmental costs of generating wealth, such as the depreciation of natural capital, are rarely taken into account?[26]

Our definition of "sustainable development" challenges the North's concept of "development," which has all too often failed to recognize the needs of specific cultures. We have seen the demise of cultural and spiritual values held by indigenous peoples, together with the loss of many of the skills needed to protect and restore specific ecosystems. These criticisms are as valid for development projects in northern Canada as they are for Amazonia. To stigmatize a society as "underdeveloped" implies there is only one acceptable route of societal evolution as demonstrated by the North's industrial countries. To define "development" in terms of the North's nomenclature is to accept uncritically an expansionist world-view which is creating planetary ecological havoc.

To be sustainable, development must include a reassessment of our current concepts of international aid. Many of the "poor" areas of the South are richest in terms of their natural biological and environmental diversity, as well as in their skills for managing their ecosystems. So instead of regarding "foreign" aid as unidirectional, new strategies for sustainability should be conceived in terms of "mutual" aid, with the appropriate transfer of environmental methods and ethics, social relationships, and technologies being recognized as bilateral of a North–South axis.

Together with other countries from the South and North, Canada might initiate and sponsor at the UN and other international forums strategies for mutual assistance. These could be integrated with an ongoing dialogue about social, economic, and environmental sustainability on a region-by-region basis. Moreover, with the world's annual arms bill amounting to about $1 trillion, Ottawa should commit itself to reformulating security policies, recognizing that mutual global security must be predicated on progressive disarmament coupled with strategies for the South's sustainable development and environmental protection.[27] Such an approach is consonant with the Report of the Pearson Commission in 1971 that "peace can only be ensured by overcoming world hunger, mass misery, and the vast disparities between rich and poor."[28]

Yet these recommended strategies to help the South move towards parity with the North presuppose, in line with the Brundtland Report's thesis, that current "mutual vulnerability" can be transformed into "mutual sustainable development." But we have had to challenge the feasibility of that thesis. While the Report alerts us to the dangers of continued unsustainable activities, it is vague and factually imprecise: its advocacy of a five- to tenfold increase in economic activity cannot be substantiated on the basis of existing evidence. Meanwhile, specific remedial action has not yet begun.

As a preliminary step, we submit our "modest proposal" (though not akin to Jonathan Swift's solution for Ireland's poor children) for a concrete investigation of specific issues. We propose as the time frame for this study the next six decades inasmuch as UN projections indicate that by the year 2050 global population could almost double to 10 billion.

1 When the IDRC (International Development Research Centre) population clock started on 16 January 1987, the planet's arable land was 1.57 billion hectares; on 23 April 1991 it had declined to 1.56 billion. It is realistic to extrapolate this rate of decline to 2050, and, if so, what will then be the population/arable land ratio as compared with today? How would this new ratio break down in terms of specific regions in the North and in the South?
2 What will be the food requirements for these regions in 2050? Can they be met from regional sources? Conversely, at current rates of agricultural production, what are the prospects of endemic hunger?
3 What will be the energy requirements to meet both population growth (as of 2050) and the South's GNP goals? Recalling the Brundtland Report's estimate that uniform worldwide energy consumption at current industrial country levels could amount to 55 TW-yr (compared with 10 TW-yr in 1980) when global population reaches 8.2 billion in 2025, what will be the consumption a quarter of a century later, when upwards of another 2 billion people have been added? What energy sources, both renewable and nonrenewable, would be required to meet such consumption? To what extent could this exponential energy increase have an impact on the atmosphere and influence climate change?
4 Apropos of environmental degradation, on the basis of current economic and social practices what might we expect as regards continued species impoverishment, the future loss of pharmaceutical products, and toxicological damage by the middle of the next century?

5 The need for a reliable quantitative base on which to make sensible forecasts includes economic projections. We may take the Brundtland Report's forecast of a five- to tenfold increase in global economic activity as an example. What kind of resource use would this require? Given resource and environmental constraints, is there a point where the North will have to cut back on its own economic growth to enable the South to catch up?
6 Political scientists and international jurists have also a role to play, for example as regards the traditional concept of national sovereignty in an increasingly interdependent global community. By the terms of chapter 7 of its charter, the UN can override national sovereignty to ensure political security. We need now to develop the concept of international environmental security, such as by codifying the Declaration of Environmental Rights into a Covenant and making it enforceable.

When we set our modern period of technological and economic expansion – coterminous with Queen's existence – within the millennia of global human society, we have to ask whether it represents an historical aberration. Previous societies evolved within an overall pattern of initial positive feedback processes, marked by incremental growth in population and resource usage, to be succeeded by moves towards a steady state. Heretofore, too, all societies managed to exist in an overall balance with the geosphere and biosphere. All relevant studies forecast that in the decades ahead the world's population and resource consumption will grow to unprecedented levels, accompanied by massive ecological degradation. Yet the Brundland Report's members and economists in general believe that it will still be possible to have a sustainable global environment with an economy five to ten times larger than the one whose activities occasioned the Planet Earth symposium. The advocates of open-ended expansionism are going to have a hard time proving that this goal is attainable.

Let us be clear about "growth." We shall need economic growth – *controlled* growth – for two basic reasons: to match the planet's demographic growth in the coming century, and to redress as far as possible the wretched disparities between North and South. But growth must be environmentally sustainable, and this imperative will require the North to cut back drastically on its percentage of planetary resource consumption.

We need to get our priorities straight. The geosphere and biosphere existed for billions of years before our species made its appearance. The planet can exist without us – very much better without us. But the converse is not true. To continue to survive, we have no alternative but to adapt our behaviour and institutions to the imperatives of environmental sustainability. Here we agree with the Brundtland Report in perhaps its most crucial conclusion. It will be excruciatingly difficult to make the required shift to both safeguard the environment and move towards parity between North and South. All such efforts must involve an unprecedented amount of political will. Do we have it?

NOTES

1 See *Energy and Power: A Scientific American Book*. San Francisco: W.H. Freeman, 1971, especially W.B. Kemp, "The flow of energy in a hunting society," R.A. Rappaport, "The flow of energy in a agricultural society," and E. Cook, "The flow of energy in an industrial society."

2 Howard, R. 1991. "World population expected to double in 60 years." *Globe and Mail*, Toronto, 14 May.

3 Head, I.L. 1991. *A World Turned Upside Down*. Dr Leonard S Klink Lecture, Agricultural Institute of Canada, 17.

4 The World Commission on Environment and Development 1987. *Our Common Future*. (Cited hereafter as Bruntland Report.) New York: Oxford University Press, ix.

5 Clow, M.J.L. 1990 "Sustainable development won't be enough." *Policy Options* 11(9): 7.

6 Bruntland Report, 43.

7 Rees, W.E. 1990. *Sustainable Development and the Biosphere: Concepts and Principles*. Teilhard Studies No. 23. Chambersburg, Pa.: Anima Books, 9.

8 Vitousek, P., et al 1986. Human appropriation of the products of photosynthesis. *Bioscience* 36: 368–74; quoted in Daly, H.E., Cobb, J.B., Jr. 1989, *For the Common Good: Redirecting the Economy Toward Community, the Environment, and a Sustainable Future*. Boston: Beacon Press, 143.

9 Bruntland Report, 213. It cites the UN Industrial Development Organization (UNIDO) estimate that world industrial output would have to be increased by a factor of 2.6 if consumption of manufactured goods in developing countries were to be raised to current industrial country levels. See UNIDO 1985. *Industry in the 1980s: Structural Change and Interdependence*. New York.

10 Bruntland Report, 168.

11 Ibid., 169–70.

12 Ibid., 169.

13 Ibid., 4.

14 This represents the wood that has been scaled, but not the timber left in the field. Meanwhile, in other parts of its environment, Canada lost 14 thousand square kilometres of farmland, due to urbanization, soil

degeneration, and erosion. In the past fifty years, 30–40 per cent of the organic matter in eastern Canada and as much as 50 per cent in the prairies have been lost. Yet less than 5 per cent of Canada's land area has both the soil and climate for sustained agriculture (while less than half of one per cent is considered "prime" agricultural land).

15 Inland Waters Directorate, Environment Canada, *Water and the Canadian Economy*; see Tables 2-1, 3-1, 3-2, appended in Foster, H.D. and Sewell, W.R.D. 1981. *Water: The Emerging Crisis in Canada.* Canadian Institute for Economic Policy. Toronto: James Lorimer.

16 Foster and Sewell, note 15 above, 17–19. For their analysis of the "myth of superabundance" see chapter 2.

17 The Council of Resource and Environment Ministers, *Report of the National Task Force on Environment and Economy,* 24 September 1987, 3.

18 Rees, W.E. 1988. Sustainable development: economic myths and ecological realities. *The Trumpeter: Journal of Ecosophy* 5(4): 135.

19 For an analysis of the Canadian federal debt and its long-term implications, see Taylor, A.M. 1990. "Here's how to get out of the budget mess," *Policy Options* 11(9): 9–13.

20 Investment Dealers Association of Canada, *Current Economic and Financial Indicators,* June 1993. A subsequent study, by the Fraser Institute, paints a still more alarming fiscal picture of the country's prospects. It measures total federal/provincial debt by including major components of public sector debt; these include unfunded liabilities of the Canada and Quebec pension plans, those of provincial and local government pensions, and those for employees of crown corporations; aboriginal land claims; and such other contingent liabilities as government guarantees to business, student loans, and megaprojects such as Hibernia. It estimates that the total all-government net debt was $1.408 trillion in 1994, resulting in a net debt-to-GDP ratio of 198 per cent. See Robin Richardson, "Inside Canada's government debt problem and the way out," *Fraser Forum,* the Fraser Institute, Vancouver, May 1994, 24–25.

21 Richardson 1994, 48.

22 Statistics Canada, National Income and Expenditure Accounts: Annual Estimates 1926–1986. *Canadian Statistical Review,* 1961–1989.

23 "Carrying capacity" may be defined as "the level of human activity (including population dynamics and economic activity) which a region's environmental systems can sustain (including consideration of import and export of resources and waste residuals) at acceptable 'quality-of-life' levels." This definition has been adapted from one used by Richard Mabbutt 1985 in *Managing Community Carrying Capacity and Quality of Life: The Boise Future Foundation Approach,* Boise, Idaho.

24 Prescott-Allen, R. 1991. *Clayoquot Sound Sustainable Development Task Force.* Report to the Minister of Environment and the Minister of

Regional and Economic Development, British Columbia Government, 31 January.

25 Brundtland Report, 23.

26 Brown, L. 1990. The illusion of progress. In *State of the world 1990.* New York: Norton, 8.

27 Roche, D. 1989. *Building Global Security: Agenda for the 1990s.* Toronto: NC Press, 56.

28 The Pearson Commission also set an aid target of 0.7 per cent of developed countries' GNP to be transferred to developing countries as official development assistance. Only Sweden, Norway, Denmark, and the Netherlands met this target, the average for donor countries remaining at about half that figure.

3 Changes in Climates of the Past: Lessons for the Future

MICHAEL B. McELROY

INTRODUCTION

The composition of the atmosphere is changing at an impressive rate at the current time, due largely to the effects of human activity. The abundance of CO_2 in the pre-industrial era was about 280 parts per million by volume (ppm); today it has risen to almost 350 ppm. The level of methane (CH_4)has increased over the same period by close to a factor of 3, from about 650 parts per billion (ppb) to almost 1800 ppb. A significant increase is observed also in the concentration of nitrous oxide (N_2O), and there are gases in the atmosphere, the industrial chloro-fluorocarbons (CFCs) for example, for which there are no natural analogues. Rising levels of CFCs and brominated analogues are responsible for significant reductions in the abundance of stratospheric ozone (O_3), most notably in polar regions, especially over Antarctica in spring. The changes observed in the composition of the atmosphere are due to a combination of influences: the use of fossil fuel (especially important for CO_2 and possibly for N_2O); deforestation (important for CO_2 and potentially for CH_4 and N_2O); agricultural practices (influencing CH_4, N_2O and possibly CO_2); and chemical industry (the dominant source of CFCs). Ultimately the changes in composition are linked to growth in the human population since the Industrial Revolution, facilitated in large measure by technology fueled by inexpensive sources of energy.

The changes in CO_2, CH_4, N_2O, O_3, and CFCs are important in that these gases, together with H_2O (gas, solid and liquid), provide the dominant contributions to the terrestrial greenhouse. The concept of the greenhouse is relatively straightforward and noncontroversial. In the absence of an atmosphere, the temperature of Earth would be determined by the albedo (reflectivity) of the surface and by the requirement that energy absorbed from the sun be balanced by radiation (mainly infrared) emitted into space. The average temperature of Earth in this case, assuming an albedo similar to today's (0.33), would be about 20°C below the freezing point of H_2O. Greenhouse gases and clouds trap a portion of the energy radiated by the surface, inhibiting otherwise direct transmission of radiant energy into space. A fraction of the energy intercepted by the atmosphere is returned to the surface, supplementing energy absorbed directly from the sun. The additional energy supplied by the atmosphere, according to the simplest energy-balance greenhouse model, is responsible for an increase in surface temperature of about 35°C and plays an important role in maintaining the surface of Earth at a relatively comfortable global average value of about 15°C.

The actual situation is more complicated. The energy balance of the surface is maintained not simply by sources and sinks of radiant energy: significant quantities of energy are removed from the surface by conduction and by the evaporation of H_2O. The energy lost from the surface by evaporation is recovered in the atmosphere as H_2O changes phase, and returns to the surface as rain or snow. The evaporation–condensation cycle provides an important, and often dominant, means for transfer of energy from the surface to the atmosphere, and from one region to another. In addition, water, both in vapour and in condensed (cloud) form, plays an essential role in the radiative budget of the atmosphere. As vapour, it represents the most important greenhouse agent, regulating to a large extent the transfer of infrared energy between the surface, the atmosphere, and space. In condensed form, it has a dual role. Clouds affect the albedo of Earth, limiting in many regions the flux of visible solar radiation that penetrates to the surface. They influence also the transfer of infrared radiation and in this sense must be considered an important component of the greenhouse. The effect of water on the energetics of the surface–atmosphere system depends on the distribution of vapour and cloud with altitude; on temperature; and on the optical properties of clouds. These properties are a sensitive function of the phase and size distribution of particles composing the cloud, as determined by, among other factors, the nature and supply of nuclei for condensation and fine details of the thermal and dynamical environment. An accurate, first principles treatment of the hydrologic cycle is beyond the scope of even the most powerful computer models available today, and this situation is not likely to change significantly in the foreseeable future.

The immediate challenge is to forecast the possible effects of the changing composition of the atmosphere on climate, or at least to estimate the potential scale of the impact. It is a daunting task. The effects on climate of, for example, an increase in the abundance of CO_2 involve a complex series of feedbacks. Feedbacks can be either positive or negative; that is to say, they can either amplify the initial change in climate or they can serve, partly at least, to counteract it.

This chapter is concerned with lessons that may be drawn from studies of the past relevant to our need to predict the future. The emphasis is on processes and feedbacks. Particular attention will be directed to hints or clues indicating that certain feedbacks are potentially more significant than is suggested by their treatment in contemporary models. I begin with a survey of climate over the past several million years (myr), focusing on the puzzle posed by the relative dominance of a 100-thousand-year (100 kyr) period in the spectrum of climate variability over the past 0.5 myr, as contrasted with the comparative absence of this signal in the earlier record. I examine the possible role of changes in the concentration of CO_2 as a factor modulating the transition from glacial to interglacial time, suggesting that CO_2 may have had a significant influence on climate of the recent past, and speculating that the absence of the 100-kyr signal in the climate record prior to 1 myr before present (BP) could reflect a higher concentration of CO_2 at that time. I continue with a discussion of the equable climates of the Cretaceous (144 to 65 myr BP) and Eocene (58 to 37 myr BP), summarizing arguments by Farrell[1] that warm climates distinguishing these epochs may have been associated with a latitudinal expansion of the symmetric (Hadley) circulation. Here again CO_2 may be implicated. A long-term, though not necessarily monotonic, decline in the concentration of CO_2 from the Cretaceous to the Holocene (the present geological epoch) could have had an important influence on the evolution of climate over the past 140 myr. I conclude with a discussion of climate change for the past several centuries, focusing on the Little Ice Age that ended in the latter half of the nineteenth century. The Little Ice Age may be considered an example of a natural fluctuation of the climate system; there is no indication of significant contemporaneous variation in the concentration of greenhouse gases. A key question is whether the end of the Little Ice Age was induced by the initial rise in the concentration of industrially derived greenhouse gases. The answer to this question could affect conclusions concerning the sensitivity of climate to the present-day increase in the concentrations of CO_2, CH_4, N_2O, and CFCs.

THE PAST 5 MILLION YEARS

Analysis of the isotopic composition of the shells of organisms preserved in the sediments of the deep sea provides a remarkable record of the changes in climate over the past 5 myr. Specifically, the relative abundance of the primary oxygen isotopes, ^{18}O and ^{16}O, expressed in terms of a quantity $\delta^{18}O$, provides a proxy for the changing abundance of water stored in continental ice.[2] The underlying physical principle is simple: evaporation of H_2O from the ocean favours the light isotope, ^{16}O; growth of continental ice sheets is reflected in an upward drift of the value for $\delta^{18}O$ of the ocean. As the ocean becomes isotopically heavy, the change in isotopic composition of ocean water, and inferentially of the volume of continental ice, is recorded in the isotopic composition of the shells of organisms growing in the sea.

Much of the analysis of the $\delta^{18}O$ record from deep sea cores has focused on the role of seasonal changes in insolation in modulating the growth and recession of continental ice sheets. The idea, attributed to Milankovitch,[3] is that expansion of continental ice sheets is favoured by relatively low summer insolation. Conversely, the ice sheets tend to recede when summer insolation is high. Seasonal changes in insolation are associated primarily with variations in the orbital parameters of Earth. Most important are changes induced by the precession of the equinoxes (winter in the Northern Hemisphere coincides today with the time when the Earth–Sun distance is close to a minimum; precession causes the position of Earth on its orbit at any particular season to migrate around the orbital path, with an associated period of about 21 kyr) and by fluctuations in the obliquity of the rotational axis (the axis is inclined with respect to the ecliptic at the present time by about 23.5°; the angle of inclination undergoes a regular oscillation with an amplitude of 3° and a period of 41 kyr), with smaller, more slowly varying, changes associated with variations in the ellipticity of the orbit. (Fourier analysis of the change in ellipticity indicates significant power in the spectrum at a period of about 100 kyr.) The variation of insolation at 60° N, calculated using formulae presented by Berger,[4] is illustrated in Figure 3.1. The change in $\delta^{18}O$ obtained from a composite of results from deep-sea cores is shown in Figure 3.2. As an aid to direct comparison, results for insolation and $\delta^{18}O$ are superimposed in Figure 3.3.

The pattern in Figure 3.3 shows clear evidence for fluctuations in climate connected with orbital forcing at frequencies associated with the changes in precession and obliquity. The evidence for forcing induced by variations in ellipticity is more ambiguous and difficult to explain. Changes in insolation caused by variations in ellipticity

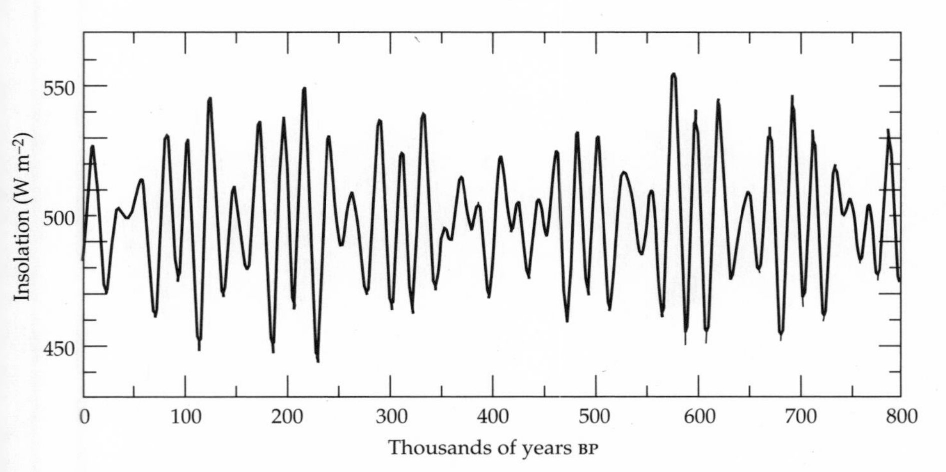

Figure 3.1
Variation of daily insolation at 60° N for summer solstice over the past 800,000 years, calculated using the formulation given by Berger (1978).

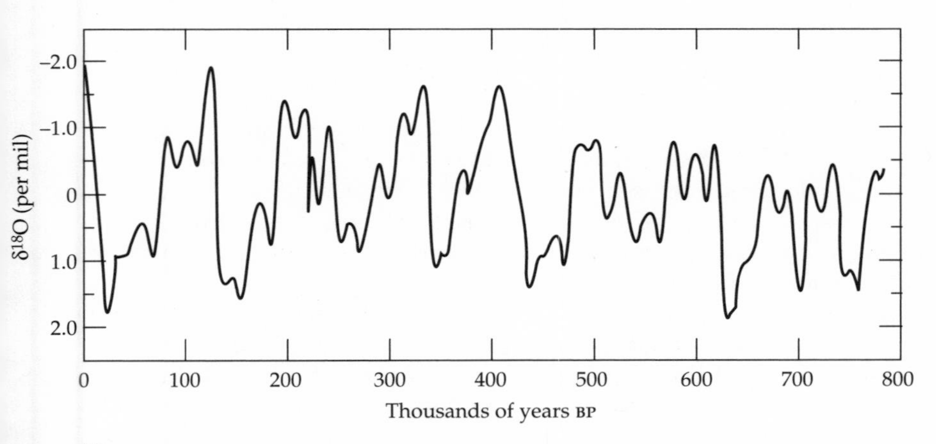

Figure 3.2
Variation of $\delta^{18}O$ of shallow-dwelling planktonic foraminifera from five deep-sea cores, plotted on the SPECMAP time-scale, over the past 800,000 years. (Imbrie et al. 1984. Reproduced by permission of Kluwer Academic Publishers.)

are small compared with the changes due to precession and obliquity. It is difficult to account for the dominance of the 100-kyr signal in the $\delta^{18}O$ record in terms of a simple response to forcing related to changes in high-latitude summer insolation. A variety of explanations has been advanced to resolve this dilemma. Two specific models have been proposed: one suggests that the 100-kyr signal is related to the dynamics of the ice sheet;[5] the other attributes a major portion

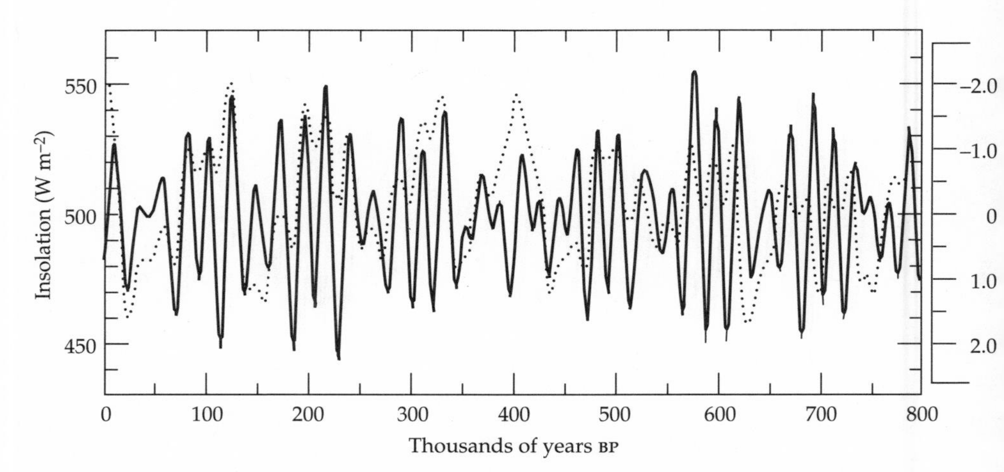

Figure 3.3
Superimposition of insolation at 60° N for summer solstice (solid line, from Figure 3.1) and the $\delta^{18}O$ record (dotted line, from Figure 3.2).

of the recent climate signal to forcing by CO_2.[6] Difficulties with both explanations are discussed below. A conceptual discussion of the problem is presented by Saltzman.[7]

The ice dynamic model suggests that the collapse of the ice sheet marking the end of a major (100 kyr) glacial epoch is triggered by the insolation-driven retreat of ice into a surface depression created by isostatic sinking of the crust beneath the ice sheet as it reaches its maximum thickness and greatest southern extent. The most elaborate treatment of the dynamic evolution of the ice sheet[8] provides a satisfactory description of the changes in $\delta^{18}O$ observed over the past 0.5 myr or so. It accounts, in particular, for the rapid demise of ice sheets observed during terminations, a phenomenon attributed to a delay in isostatic rebound compared with the time required for ablation and melting of the ice sheet at its southern margin. The model adopts an empirical formula to define the rate at which snow accumulates as a function of height and latitude and accounts for a strong nonlinear dependence of the rate of ablation on the height of the ice sheet. The model is driven by realistic, time-varying, summer insolation. The 100-kyr period results from a particular choice of parameters defining rates for accumulation and ablation.

There are several problems with the ice dynamic model. First, as noted by Broecker and Denton,[9] it fails to account for the near-synchroneity of warming observed in both hemispheres during terminations: a change in the heat budget at high latitudes in one hemisphere is not readily transmitted by the atmosphere to the opposite hemisphere, as shown by the general circulation model (GCM) study of Manabe and Broccoli.[10] Second, the 100–kyr signal is weak, or absent, from the climate record prior to about 1 myr BP.[11] A period of 41 kyr is the dominant feature of the climate in the early Pleistocene (1.66 to 0.73 myr BP) and for the duration of the Pliocene (5.1 to 1.66 myr BP). A low-frequency signal sets in gradually at about 1 myr BP and evolves steadily, developing into the 100 kyr feature by about 0.6 myr BP.[12] The change in the pattern of climate variability since 5.1 myr BP would appear to have required a corresponding change in forcing. The shift could be attributed to a secular, though not necessarily monotonic, decline in CO_2 over the interval 5.1 to 0.6 myr BP.

The CO_2 hypothesis provides a promising explanation of climate variability. GCMs suggest that the difference in the concentration of CO_2 between glacial and interglacial time, 200 as compared with 280 ppm,[13] could have an important influence for both global and regional climate.[14] Particularly pertinent in this context is a study by Broccoli and Manabe.[15] They used a GCM to simulate the atmosphere with a spatial resolution of 4.5° in latitude by 7.5° in longitude coupled to a simple model of the ocean mixed layer. (The model allowed for storage of heat by the surficial ocean but did not attempt to account for its spatial redistribution by ocean circulation.) The distribution of clouds was prescribed on the basis of contemporary observations and the model was shown to provide a satisfactory description of the present climate. Introduction of an ice sheet similar to that observed at the last glacial maximum (LGM) lowered the average sea surface temperature (SST) by 0.8°C, with most of the drop predicted for the Northern Hemisphere (1.6°C) and a much smaller impact in the south (0.2°C). In contrast, the reconstruction of SST by CLIMAP[16] for 18 kyr BP indicates declines in the average values of SST for the Northern Hemisphere, the Southern Hemisphere and for the globe of 1.9, 1.3 and 1.6°C respectively. It is clear that direct effects of continental ice sheets are confined largely to the Northern Hemisphere. Allowing for a drop in CO_2 from 300 to 200 ppm improves the agreement between model and observation: SST of the Northern Hemisphere, accounting for the change both in ice and CO_2, declines in the model by 2.6°C while the corresponding reductions for the Southern Hemisphere and for the globe are 1.5 and 1.9°C respectively, in much better agreement with observation.

Additional evidence for a connection between changes in CO_2 and associated variations in climate comes from analysis of the abundance of deuterium (D) and CO_2 from the Vostok ice core.[17] The abundance of D relative to H in the ice, expressed by the quantity δD,

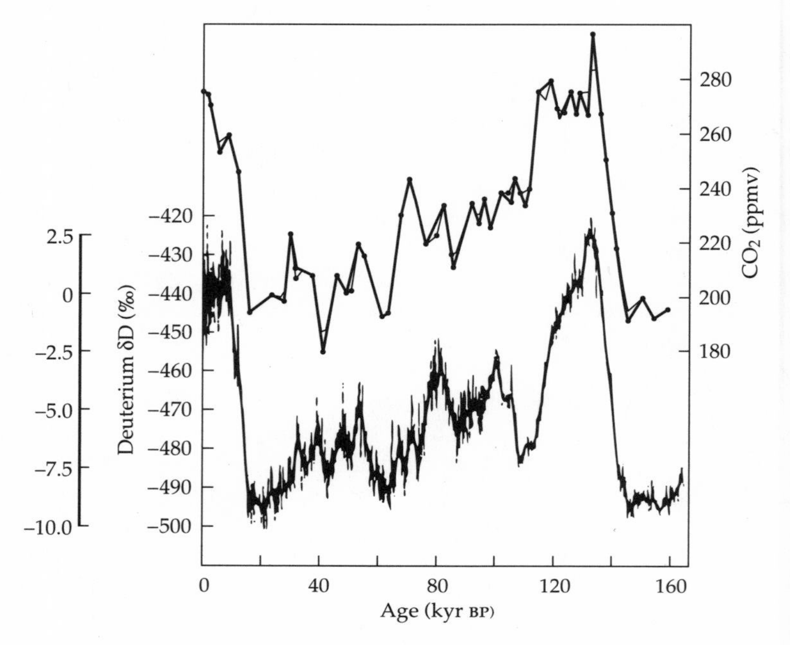

Figure 3.4
Variation of the concentration of the CO_2 (upper curve) and the relative abundance of deuterium (lower curve) over the past 160,000 years, from the Vostok ice core (East Antarctica). Also shown is the temperature change associated with the variation of deuterium. (Jouzel et al. 1987; Barnola et al. 1987 [Reproduced with permission from *Nature*, ©1987, Macmillan Magazines Ltd.]).

provides a proxy for the annual mean local temperature at the time of precipitation.[18] The variation of temperature inferred for the past 160 kyr in this fashion is shown as a function of the concentration of CO_2 measured in the same core (Figure 3.4). There are obvious similarities in the two records. In particular, the rapid increases in temperature, by about 10°C at the end of the last and penultimate ice ages (at about 15 and 140 kyr) are associated with comparably rapid increases in CO_2 (from about 200 to 280 ppm). The drop in temperature over the past ice age is associated with a compatible decline in CO_2, although in this case the fall in temperature appears to lead the decline in CO_2. This is not surprising. The change in CO_2 from glacial to interglacial time is undoubtedly associated with release of CO_2 from the ocean.[19] Uptake of CO_2 by the ocean is limited by the supply of CO_3^{--} to surface waters, while there is no comparable limitation on release. Genthon et al.[20] sought to relate the climate signal recorded by the Vostok core to a combination of forcing due to CO_2, expressed in terms of an empirical formula taken from Hansen et al.,[21] amplified by changes in ice volume expressed either by the variation of summer insolation at high latitudes of the Northern Hemisphere (the Milankovitch effect) or by the $\delta^{18}O$ record from deep-sea cores,[22] combined with a regional influence simulated either by the temporal variability of insolation in local summer (November) at 60° S or by the annually averaged intensity of insolation at 78° S (the location of Vostok). In this manner they obtained a statistically excellent fit to the climate record (r^2 typically about 0.9, where r defines the correlation between calculated and observed values of ΔT). A major portion of the climate forcing, as much as 85 per cent, was attributed to changes in CO_2.

In summary, it may be concluded from this survey of variations in climate over the past several million years (1) that changes in CO_2 had an important influence on climate, at least for the past 160 kyr; (2) that changes in the rhythm of climate observed about 1 myr BP (from a dominant period in earlier years of about 40 kyr to a more recent value of close to 100 kyr) may have been due to a temporal decline in the average baseline level of CO_2, associated perhaps with an increase in the rate of weathering; and (3) that, while variability in CO_2 on time-scales of about 100 kyr is controlled mainly by changes in the circulation of the ocean, the ocean–atmosphere system is coupled, with CO_2 having a significant influence on the circulation of the ocean.

EQUABLE CLIMATES OF THE CRETACEOUS AND EOCENE

We turn our attention now from the ice-dominated world of the past 5 million years to the relatively warm climates of the Cretaceous (144 to 65 myr BP) and Eocene (58 to 37 myr BP). Measurements of the isotopic composition of oxygen in the shells of benthic organisms preserved in marine sediments provide a useful record of temporal changes in the temperature of the deep ocean. The data, summarized in Figure 3.5 indicate a decline in deep water temperatures from about 18°C in the Cretaceous to near 0°C today. While interpretation of the isotopic measurements is not unambiguous – inferences on temperature depend, for example, on assumptions made with respect to salinity, for which there are no independent data – it is clear that the pattern exhibited in Figure 3.5 offers strong support for the view that temperatures at high latitude were much higher during the Cretaceous and Eocene than they are today.

The terrestrial flora and fauna tell a similar tale. Frost-intolerant vegetation was common in Spitsbergen (palaeolatitude 79° N) during

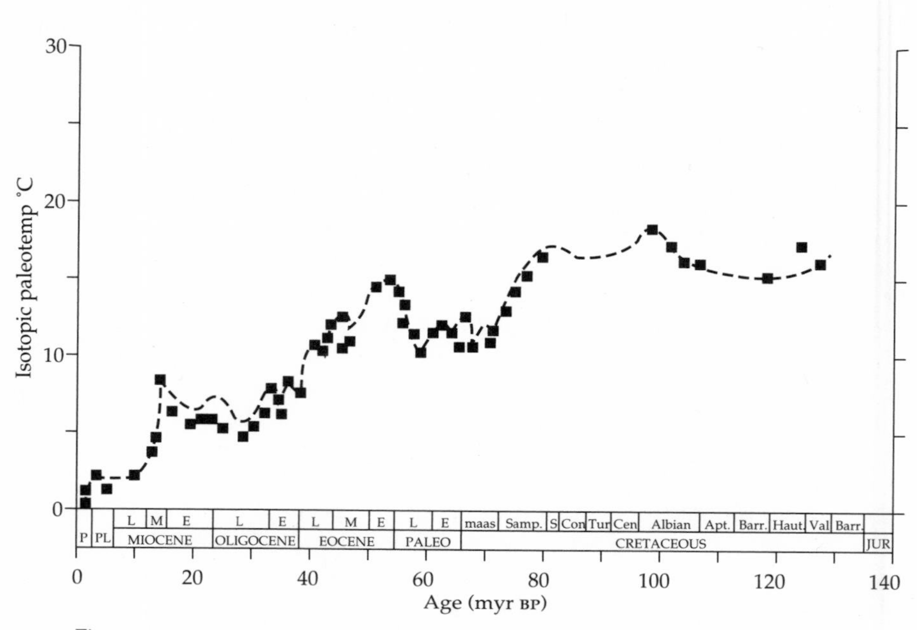

Figure 3.5
Reconstructed temperature of the tropical Pacific ocean over the past 140 myr for bottom-water, based on benthic foraminifera. (Reproduced with permission from Douglas, R.G., Woodruff, F. 1981. Deep sea benthic foraminifera. In *The Sea*, vol. 7, ed. C. Emiliani. New York: Wiley-Interscience, 1233–1327.)

the Eocene.[23] Alligators[24] and flying lemurs[25] are observed in deposits of comparable age from Ellesmere Island (palaeolatitude 78° N), while data from central Asia (palaeolatitude 60° N) indicate the presence of palm trees in this region during the Cretaceous.[26] As recent as 5 myr BP, the forest–tundra boundary extended to latitudes as high as 82° N, some 2500 km north of its present location, occupying regions of Greenland now perpetually covered in ice.[27]

The presence of flora incapable of surviving even occasional frost in the interior of continents at high latitude convinced Farrell[28] that equable climates of the past should reflect a major expansion of the tropical Hadley circulation. Heat is transported from low to high latitudes by the Hadley circulation, mainly in the upper troposphere, with a return flow near the surface. Air parcels moving to higher latitudes are deflected to the east by the Coriolis force; this tends to limit the efficiency with which heat is redistributed by meridional motion. More efficient exchange of angular momentum, either with the surface or with higher latitudes, would allow the Hadley circulation to expand in latitude. Other factors contributing to the efficiency with which heat is transported meridionally by the Hadley circulation include the height of the tropopause, the vertical gradient of temperature in the troposphere, the time-scale for radiative relaxation, and the horizontal gradient of temperature appropriate for radiative equilibrium. A high tropopause provides a deeper troposphere for transport of heat. The more stable the atmosphere, the larger the relative heat content of the air transported in the upper branch of the circulation cell. The longer the time-scale for radiative relaxation, the more effective the role of the circulation in transporting heat. The smaller the latitudinal gradient of the temperature of the atmosphere in radiative equilibrium, the lower the demand imposed on the circulation to maintain a specified gradient of temperature. The efficiency of the Hadley circulation is measured in terms of a dimensionless number, Γ, given by Farrell in the formula

$$\Gamma = \frac{S\tau_R}{\delta_H \tau_A},$$

where S defines the mean static stability of the troposphere, τ_R denotes the time constant for radiative relaxation, δ_H specifies the fractional change of the radiative equilibrium temperature with latitude, and τ_A refers to the time constant for dissipation of angular momentum. Farrell estimates a value for Γ in the present climate of 0.5 and concludes that an equable climate would require an increase in Γ by about a factor of 8.

There are reasons to believe that CO_2 was relatively abundant during the Cretaceous and Eocene. Arthur et al.,[29] interpreting measurements of the isotopic composition ($\delta^{13}C$) of organic material from phytoplankton preserved in Cretaceous sediments, concluded that the level of CO_2 was four to twelve times higher during the Cretaceous than it is today.[30] The concentration of CO_2 declined during the Cenozoic (65 to 0 myr BP), with an abrupt decrease to less than twice the present abundance during the early to middle Miocene (24 to 5 myr BP). It is tempting to associate the rapid decline in CO_2 in the Miocene with the drop in the temperature of deep water indicated for about the same time by the data in Figure 3.5 and to attribute the eightfold increase in Γ required to account for equable climates to feedbacks associated with a higher level of CO_2.

Warmer temperatures during the Cretaceous and Eocene would have permitted the atmosphere to hold larger quantities of H_2O. Release of latent heat associated with an enhanced burden of H_2O

would have contributed to more effective redistribution of energy in the vertical, resulting in a larger value of S. Increased convective activity would have promoted more effective coupling of angular momentum between the surface and the upper troposphere, resulting in a lower value for τ_A. In addition, increased burdens of CO_2 and H_2O could have contributed to a decrease in the value for τ_R. These factors could have combined to provide the necessary change in Γ. A warmer climate could have been facilitated further by an increase in the depth of the troposphere, as discussed by Farrell.[31]

Recognition that the isotopic composition of the organic fraction of phytoplankton contains information on the abundance of CO_2 dissolved in sea water[32] has led to a powerful new technique for the study of past variations of CO_2. Further advances may be anticipated from measurements of specific molecular compounds in phytoplankton[33] and from studies of the isotopic composition of carbonates in palaeosols.[34] As discussed here, preliminary results from analyses of bulk organic materials in phytoplankton[35] indicate that the abundance of CO_2 was much higher in the Cretaceous than it is today, and that it declined steadily through the Cenozoic. It should be possible with further work to refine this analysis in order to identify times in the past when concentrations of CO_2 were comparable to values expected to develop in the future as a consequence of human activity. Studies of associated climates could provide invaluable insight into conditions anticipated in the future as a response to human activity.

CLIMATES OF THE RECENT PAST

A record of temperature change for the past 10,000 years inferred for western Norway in the region of the Jostedalsbreen ice cap using a variety of lithostratigraphic and palaeobotanical techniques[36] is presented in Figure 3.6. The data document the recovery of climate from the Younger Dryas cold epoch 10,000 years BP. Warmest temperatures over the past 10,000 years occurred during the period known as the Hypsithermal, from about 8,000 to about 5,000 years BP. The Jostedalsbreen ice cap disappeared during the Hypsithermal. Temperatures have declined more or less steadily since then, with the ice cap reappearing about 5,000 years ago, reaching its maximum extent since the Erdalen during the period known as the Little Ice Age.

Figure 3.6 includes a curve, derived using formulae presented by Berger,[37] showing the variation of insolation with time for the summer solstice at 60° N. It is intriguing to note the similarity in the behaviour

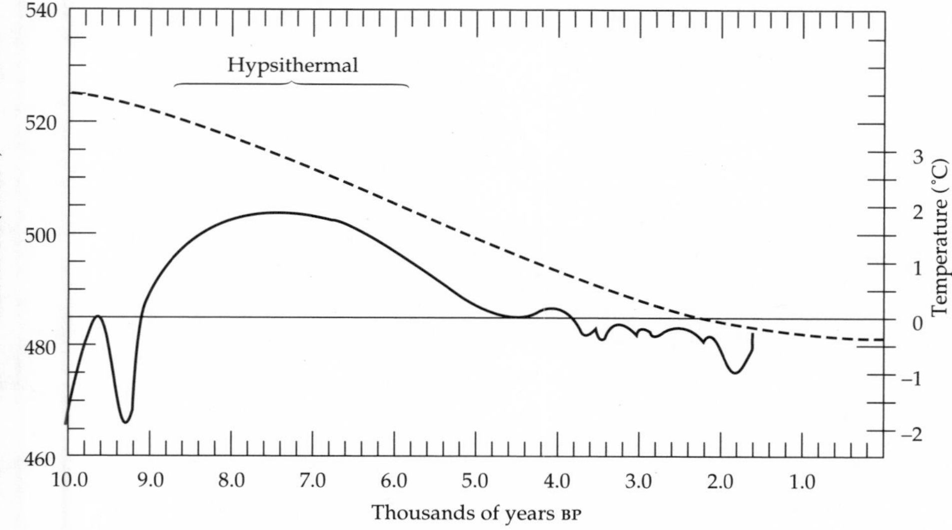

Figure 3.6
Variation of temperature (solid curve) for the Jostedalsbreen region of Norway (62° N, 7° E), based on a variety of lithostratigraphic and palaeobotanical techniques (Nesje and Kvamme 1991), and the variation of daily insolation (dashed curve) at 60° N for summer solstice (based on Berger's formulation (1978) over the past 10,000 years.

of the temperature and insolation curves for the period subsequent to about 6,500 years BP, after sea level had stabilized close to its present value. If the insolation curve is taken as a surrogate for temperature, the Little Ice Age appears as a relatively modest negative excursion with respect to the long-term trend. If physical significance is attached to the similarity between the trends in temperature and insolation after 6,500 yr BP, (and in the absence of a quantitative model this may be dubious) we must account for the quite different behaviour observed earlier. The earlier trend could reflect feedbacks associated with the demise of the global ice sheets, associated, for example, with changes in albedo and/or related changes in ocean circulation.

The Little Ice Age was a period of generally cold temperatures, specifically cold winters, lasting from about AD 1550 (or perhaps from as early as AD 1250) to about AD 1850, with extremes observed between about AD 1580 and 1700. It was marked by a significant decline in sea surface temperatures both in the North Atlantic[38] and in the North Pacific.[39] The impact appears to have been global, with important changes reported not only for Europe[40] but also for North America,[41] China,[42] Japan[43] and South America.[44] There is evidence

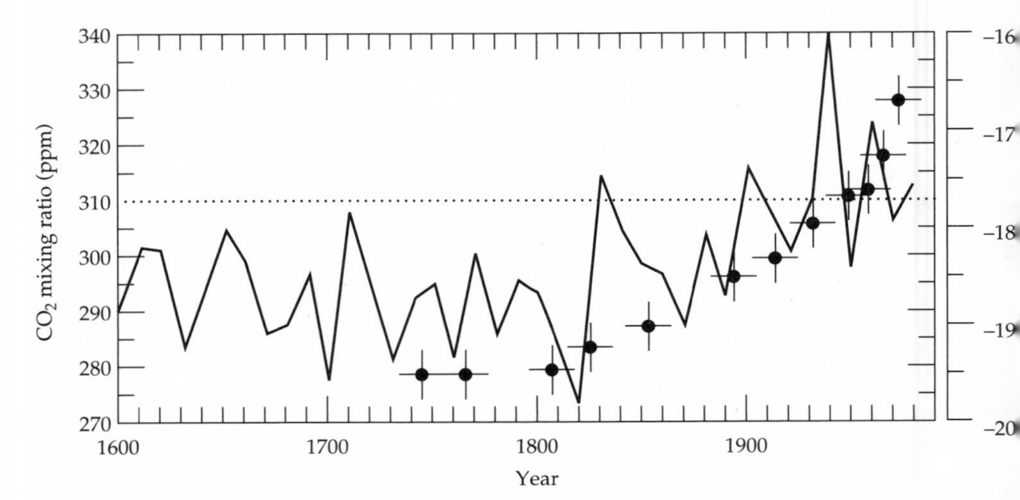

Figure 3.7
Variation of the concentration of CO_2 (data points) measured in air bubbles from an ice core obtained at Siple Station, Antarctica 76° S, 84° W (Neftel et a. 1988) and the deca-dal average value of $\delta^{18}O$ (solid line) from the summit of the Quelccaya ice core (Thompson, L.G., Mosley-Thompson, E. 1989. One-half millennia of tropical climate variability as recorded in the stratigraphy of the Quelccaya ice cap, Peru. In *Aspects of Climate Variability in the Pacific and the Western Americas*, ed. D.H. Peterson. Washington, D.C.: American Geophysical Union, 15–31).

for a worldwide advance of mountain glaciers, with snowlines declining by about 100 to 200 m at middle latitudes and by as much as 300 m in the equatorial Andes.[45]

It appears that the end of the Little Ice Age was abrupt and that it took place globally in the latter half of the nineteenth century.[46] It is unclear whether the recovery from the Little Ice Age was simply a natural response of the climate system or whether it may have been promoted in part by the postindustrial increase in the concentration of greenhouse gases. The changes are essentially synchronous, as illustrated in Figure 3.7.

Estimates for the change in temperature of the Northern Hemisphere, Southern Hemisphere, and globe are presented for the past 130 years in Figure 3.8.[47] In contrast to earlier work,[48] the study summarized here accounts for data on the time variation of sea surface temperature. Despite gaps in data coverage, especially over the ocean and at high latitudes of the Southern Hemisphere, the results in Figure 3.8 are believed to provide a reasonable (and in any event the best available) indication of the trend in global and hemispheric temperatures for the past century. The data suggest that the

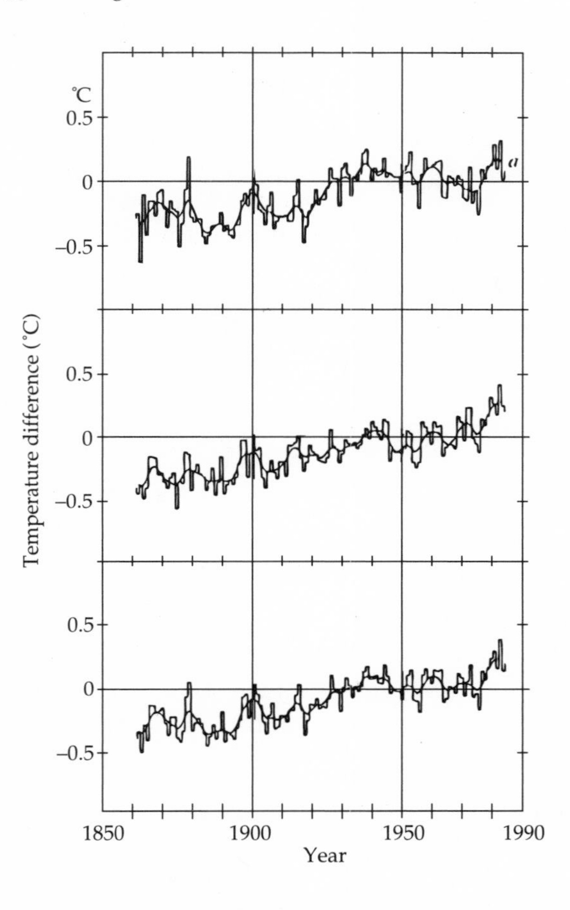

Figure 3.8
Annual temperature variations since 1861 for the global average (upper panel); the Northern Hemisphere (middle panel), and the Southern Hemisphere (lower panel), based on sea surface and land surface temperature records. The smooth curves represent ten-year Gaussian filtered values. (Jones et al. 1986. Reproduced with permission from *Nature*, ©1986, Macmillan Magazines Ltd.)

temperature of the Northern Hemisphere, and of the Southern Hemisphere and globe, has risen by about 0.5°C over the last hundred years. Much of the increase occurred recently and over the last few decades of the nineteenth century, with the earlier change associated almost certainly with the terminal stage of the Little Ice Age. The trend observed for the Southern Hemisphere is more regular than that for the north, where temperatures declined initially during the early part of the twentieth century and then again in the thirties and forties. It is tempting to attribute at least a portion of the difference between the hemispheres to an increase in cloud reflectivity[49] resulting from an enhanced burden of sulfate-based condensation nuclei formed by the oxidation of industrially derived sulphur dioxide (SO_2)[50]. Industrial sulphur could contribute also to the lack of a clearly established trend for temperatures over the continental United States;[51] indeed, there is observational evidence of an increase in cloud cover over North America during the past century, with the largest increase, observed between 1930 and 1950, roughly coincident with the period of modest cooling implied for the Northern Hemisphere as a whole in Figure 3.8.[52]

If the postulated impact of sulfur on climate is real, we are left with a bothersome dilemma. Steps taken to reduce emission of industrial SO_2 to mitigate the effects of acid rain may have an undesirable consequence: they could result in an acceleration of the warming due to greenhouse gases.

CONCLUDING REMARKS

Circumstantial evidence suggests that variations in CO_2, combined with shifts in ocean circulation, have had a significant influence on climate over much of the past 50 million years. Models imply that low levels of CO_2 (about 200 ppm) were required to account for the global nature of cooling observed during the two most recent ice ages and that changes in CO_2 (increases of about 80 ppm) may have contributed to the rapidity of warming associated with glacial terminations. It is likely that CO_2 had a similar influence during previous episodes of glaciation. A change in the rhythm of climate variability approximately one million years ago, i.e. a shift from a dominant period of about 41,000 yr to one of about 100,000 yr, may have been due to a long-term decline in the level of CO_2.

The concentration of CO_2 reflects, in the short term ($\leq$100,000 years), an equilibrium between the atmosphere and the ocean. Apportionment of CO_2 between the atmosphere and ocean depends in a complex fashion on the chemical, physical, and biological state of the sea, regulated ultimately by climate, modulated in turn by CO_2, defining a system coupled so intimately that it is difficult at any given time to isolate cause from effect. The concentration of carbon in the atmosphere–ocean system, and consequently of CO_2, can vary on longer time-scales due to changes in volcanism and tectonic activity affecting rates at which CO_2 is produced and consumed by the Earth's crust. It was suggested that equable climates of the Cretaceous and Eocene may have been maintained by higher levels of CO_2. The decline in CO_2 since the Cretaceous may be due to an increase in the rate of consumption of the gas by weathering relative to the rate at which it is produced by volcanism.[53]

The geologic evidence suggests that the climate was much warmer when the concentration of CO_2 exceeded its present value by about a factor of two. This was implied, for example, by the temporal trend in the temperature of the deep sea illustrated in Figure 3.5. It may be improper, however, to conclude that an ice-free, equable, planet is the inevitable fate of continued growth in the concentration of greenhouse gases. Feedbacks involving ice (continental and marine) play an important role in the present climate. Such feedbacks were almost certainly absent, or at least different, during the warmer climates of the early Cenozoic, when the abundance of CO_2 may have been comparable to that expected to arise over the next few decades as a consequence of human activity. More than one climate system distinguished by different modes of ocean circulation could exist in equilibrium with a particular level of CO_2.

I view with pessimism, at least for the near future, the prospects for accurate prediction of the response of climate to present and anticipated future levels of greenhouse gases. There are problems both with the atmosphere and with the ocean. The resolution of GCMS for the atmosphere is currently inadequate. Many of the centrally important physical processes, especially those relating to the hydrologic cycle, are treated in an ad hoc fashion by models, with parameterization schemes designed to reproduce features of the present climate. Even in this limited application the models are suspect, as shown by Stone and Risbey.[54] I doubt that the problems can be circumvented by resorting to faster computers which allow improved spatial and temporal resolution; the scales of importance are simply too small and the underlying physics too poorly understood. We need to develop a better understanding of the relevant processes. There is a need for simple models designed to test specific hypotheses illuminated by observation. With improved understanding, advances can be incorporated in more complex models. Only in this fashion can we hope to establish a capability for useful prediction. The acid test of models must involve a demonstration of their ability to simulate climates of the present and past for which useful data exist or for which they may be derived.

Ultimately, we need models for the coupled atmosphere–ocean system. It is likely that small-scale processes are as important for the ocean as they are for the atmosphere. We need to develop a better understanding of processes regulating the production of deep and intermediate waters. Advances in this area are likely to require major improvements in our capacity to observe the ocean. Studies of the past can be useful in focusing attention on potentially important processes.

The record of the change in climate over the past several thousand years obtained from the study of the Jostedalsbreen ice cap, combined with more specific information for the Little Ice Age, provides a useful context in which to view the possibility of contemporaneous changes relating to human activity. The difficulty associated with objective assessment of the human impact is compounded by the deficiencies of models, by the natural variability of the climate system (associated presumably with fluctuations in ocean circulation), by the evidence for long-term cooling indicated in Figure 3.6, by the relatively modest scale of the increase in temperature observed over the past century, and by the possibility that warming due to increased concentrations of greenhouse gases may be offset partially by cooling promoted by anthropogenic sources of sulphur. Under the circumstances, it is not surprising that the reactions are mixed. On the one hand, it might be argued in line with Lindzen[55] that warming over the past century has been comparatively minor, within the natural noise of the climate system, that radiative forcing due to the increase in greenhouse gases is small (about one per cent of the incident solar flux), and that models ignore feedbacks that might serve to counteract warming attributable to the direct effect of greenhouse gases (for example, a decrease in the abundance of H_2O in the upper troposphere induced by enhanced precipitation). It could be suggested indeed that warming might be advantageous; the climate of the latter half of the twentieth century is certainly more pleasant than the harsh conditions of the seventeenth century in the depths of the Little Ice Age. On the other side, one could point to the velocity of the change in atmospheric composition now underway; one could emphasize the rapidity with which climate has shifted at particular times in the past; one could highlight the fact that earlier climates were much warmer than that of today, with levels of CO_2 comparable to values expected in the near future; one could speculate on feedbacks not yet identified that could amplify the scale of warming for the next few decades; one could argue that the introduction of controls on emission of industrial SO_2 could accelerate the pace of warming; and one could point to the potential significance of social, political and environmental dislocations that could ensue if climate were to undergo a rapid shift. We are faced with a political choice: to act now to mitigate the rise in the concentration of greenhouse gases when the impact on climate is uncertain; or to wait for an improved understanding when the impact may be already significant. It is clear in either case that the stakes are high.

SUMMARY

Climate has varied over a large range in the recent history of Earth, with extremes represented by the equable environments of the Cretaceous and Eocene and the comparatively frigid conditions of the ice ages that punctuated the past few million years. It is suggested that shifts in CO_2 have played an important role in climate changes of the past. Compared to today, levels of CO_2 were 4 to 12 times higher during the warm period of the Cretaceous and about 40 per cent smaller during the last ice age. We argue that studies of past climates, by drawing attention to important processes and feedbacks, can play a valuable role in the development of credible models for the future.

ACKNOWLEDGMENTS

This work was supported by the National Science Foundation under Grant #NSF-ATM-89–21119 to Harvard University. I am indebted to Ross J. Salawitch for technical assistance in the preparation of this paper.

NOTES

1 Farrell, B.F. 1990. Equable climate dynamics. *Journal of the Atmospheric Sciences* 47: 2986–95.
2 Emiliani, C. 1955. Pleistocene temperatures. *Journal of Geology* 63: 538–78; Broecker, W.S., Van Donk, J. 1970. Insolation changes, ice volumes and the δ^{8}o record in deepsea cores. *Reviews of Geophysics and Space Physics* 8: 169–98; Shackleton, N.J. 1967. Oxygen isotope analyses and Pleistocene temperatures re-assessed. *Nature* 215: 15–17; Shackleton, N.J., Opdyke, N.D. 1973. Oxygen isotope and paleomagnetic stratigraphy of equatorial Pacific core V28–238: oxygen isotope temperatures and ice volumes on a 10^5-year and 10^6-year scale. *Quaternary Research* 3: 39–55; Hays, J.D., Imbrie, J., Shackleton, N.J. 1976. Variations in the Earth's orbit: pacemaker of the ice ages. *Science* 194: 1121–32; Imbrie, J., Hays, J.D., Martinson, D.G. et al. 1984. The orbital theory of Pleistocene climate: support from a revised chronology of the marine $\delta^{18}O$ record. In *Milankovitch and Climate*, Part 1, ed. A.L. Berger et al. Dordrecht, Holland: D. Reidel, 269–305.
3 Milankovitch, M. 1941. *Canon of Insolation and the Ice Age Problem*. Royal Serbian Academy Special Publication 133, Belgrade. (Translated by the Israel Program for Scientific Translation, Jerusalem, 1969.)
4 Berger, A.L. 1978. Long-term variations of daily insolation and Quaternary climatic changes. *Journal of the Atmospheric Sciences* 35: 2362–7.

5 Weertman, J. 1976. Milankovitch solar radiation variations and ice-age ice-sheet sizes. *Nature* 261: 17–20; Pollard, D., Ingersoll, A.D., Lockwood, J.G. 1980. Response of a zonal climate–ice sheet model to the orbital perturbations during the Quaternary ice ages. *Tellus* 32: 301–319; Oerlemans, J. 1980. Model experiments on the 100,000-yr glacial cycle. *Nature* 287: 430–32; Birchfield, G.E., Weertman, J., Lunde, A.T. 1981. A paleoclimate model of Northern Hemisphere ice sheets. *Quaternary Research* 15: 126–42; Pollard, D. 1982. A simple ice sheet model yields realistic 100 kyr glacial cycles. *Nature* 296: 334–8; Pollard, D. 1983a. A coupled climate–ice sheet model applied to the Quaternary ice ages. *Journal of Geophysical Research* 88: 7705–18; Pollard, D. 1983b. Ice-age simulations with a calving ice-sheet model. *Quaternary Research* 20: 3048; Pollard, D. 1984. Some ice-age aspects of a calving ice-sheet model. In *Milankovitch and Climate*, Part 2, ed. A.L. Berger et al. Dordrecht, Holland: D. Reidel, 541–64; Peltier, W.R., Hyde, W.T. 1984. A model of the ice age cycle. In *Milankovitch and Climate*, Part 2, 565–80; Hyde, W.T., Peltier, W.R. 1985. Sensitivity experiments with a model of the ice age cycle: the response to harmonic forcing. *Journal of the Atmospheric Sciences* 42: 2170–88; Hyde, W.T., Peltier, W.R. 1987. Sensitivity experiments with a model of the ice age cycle: the response to Milankovitch forcing. *Journal of the Atmospheric Sciences* 44: 1351–74.

6 Pisias, N.G., Shackleton, N.J. 1984. Modeling the global climate response to orbital forcing and atmospheric carbon dioxide changes. *Nature* 310: 757–9; Genthon, C., Barnola, J.M., Raynaud, D. et al 1987. Vostok ice core: climatic response to CO_2 and orbital forcing over the last climatic cycle. *Nature* 329: 414–18.

7 Saltzman, B. 1990. Three basic problems of paleoclimatic modeling: a personal perspective and review. *Climate Dynamics* 5: 67–78.

8 Peltier and Hyde 1984; Hyde and Peltier 1987.

9 Broecker, W.S., Denton, G.H. 1989. The role of ocean–atmosphere reorganizations in glacial cycles. *Geochimica et Cosmochimica Acta* 53: 2465–501.

10 Manabe, S., Broccoli, A.J. 1985. The influence of continental ice sheets on the climate of an ice age. *Journal of Geophysical Research* 90: 2167–90.

11 Joyce, J.E., Tjalsma, L.R.C., Prutzman, J.M. 1990. High resolution planktic stable isotope record and spectral analysis for the last 5.35 M.Y.: ocean drilling program site 625 northeast Gulf of Mexico. *Paleoceanography* 5: 507–29; Saltzman 1990.

12 Joyce et al. 1990.

13 Barnola, J.M., Raynaud, D., Korotkevich, Y.S., Lorius, C. 1987. Vostok ice core provides 160,000-year record of atmospheric CO_2. *Nature* 329: 408–14; Neftel, A., Oeschger, H., Staffelbach, T., Stauffer, B. 1988. CO_2 record in the Byrd ice core 50,000–5,000 years BP. *Nature* 331: 609–11.

14 Manabe and Broccoli 1985; Manabe, S., Bryan, Jr.,K. 1985. CO_2-induced change in a coupled ocean–atmosphere model and its paleoclimatic implications. *Journal of Geophysical Research* 90: 11689–707; Rind, D. 1987. Components of the ice age circulation. *Journal of Geophysical Research* 92: 4241–81; Rind, D., Peteet, D., Kukla, G. 1989. Can Milankovitch orbital variations initiate the growth of ice sheets in a general circulation model? *Journal of Geophysical Research* 94: 12851–71; Oglesby, R.J., Saltzman, B. 1990. Sensitivity of the equilibrium surface temperature of a GCM to systematic changes in atmospheric carbon dioxide. *Geophysical Research Letters* 17: 1089–92.

15 Broccoli, A.J., Manabe, S. 1987. The influence of continental ice, atmospheric CO_2, and land albedo on the climate of the last glacial maximum. *Climate Dynamics* 1: 87–99.

16 CLIMAP Project Members 1976. The surface of the ice-age Earth. *Science* 191: 1131–7; CLIMAP Project Members 1981. Seasonal reconstruction of the Earth's surface at the last glacial maximum. *Geological Society of America* Map and Chart Series, No. 36.

17 Genthon et al. 1987.

18 Jouzel, J., Lorius, C., Petit, J.R., Genthon, C., Barkov, N.I., Kotlyakov, V.M., Petrov, V.M. 1987. Vostok ice core: a continuous isotope temperature record over the last climatic cycle (160,000 years). *Nature* 329: 403–8.

19 Broecker, W.S. 1982. Glacial to interglacial changes in ocean chemistry. *Progress in Oceanography* 11: 151–97.

20 Genthon et al. 1987.

21 Hansen, J., Russell, G., Lacis, A., Fung, I., Rind, D., Stone, P. 1985. Climate response times: dependence on climate sensitivity and ocean mixing. *Science* 229: 857–9.

22 Martinson, D.G., Pisias, N.G., Hays, J.D., Imbrie, J., Moore, Jr., T.C., Shackleton, N.J. 1987. Age dating and the orbital theory of the ice ages: development of a high resolution 0 to 300,000 year chronostratigraphy. *Quaternary Research* 27: 1–29.

23 Schweitzer, H.J. 1980. Environment and climate in the early tertiary of Spitsbergen. *Palaeogeography, Palaeoclimatology, Palaeoecology* 30: 297–311.

24 Dawson, M.R., West, R.M., Langston, Jr., W., Hutchison, J.H. 1976. Paleogene terrestrial vertebrates: northernmost occurrence, Ellesmere Island, Canada. *Science* 192: 781–2.

25 McKenna, M. 1980. Eocene paleolatitude, climate, and mammals of Ellesmere Island. *Palaeogeography, Palaeoclimatology, Palaeoecology* 30: 349–62.

26 Vakhrameev, V.A. 1975. Main features of phytogeography of the globe in Jurassic and Early Cretaceous time. *Paleontological Journal* 2: 123–33.

27 Funder, S., Abrahamsen, N., Bennike, O., Feyling-Hanssen, R.W. 1985. Forested Arctic: evidence from North Greenland. *Geology* 13: 542–6;

Carter, L.D., Brigham-Grette, J., Marincovich, Jr., L., Pease, V.L., Hillhouse, J.W. 1986. Late Cenozoic Arctic Ocean sea ice and terrestrial paleoclimate. *Geology* 14: 675–8.

28 Farrell 1990.

29 Arthur, M.A., Hinga, K.R., Pilson, M.E.Q., Whitaker, D., Allard, D. 1991. Estimates of pCO_2 for the last 120 myr based on the $\delta^{13}C$ of marine phytoplankton organic matter. *Elsevier Oceanography Series* 72: 166.

30 Rau G.H., Takahashi, T., Des Marais, D.J. 1989. Latitudinal variations in plankton $\delta^{13}C$: implications for CO_2 and productivity in past oceans. *Nature* 341: 516–18.

31 Farrell 1990.

32 Rau et al 1989; Popp, B.N., Takigiku, R., Hayes, J.M., Louda, J.W., Baker, E.W. 1989. The post-Paleozoic chronology and mechanism of ^{13}C depletion in primary source organic matter. *American Journal of Science* 289: 436–54.

33 Jasper, J.P., Hayes, J.M. 1990. A carbon isotope record of CO_2 levels during the late Quaternary. *Nature* 347: 462–4.

34 Cerling, T.E. 1991. Stable isotopic constraints on atmospheric pCO_2 from paleosols. *Elsevier Oceanography Series* 72: 166.

35 Arthur et al. 1991.

36 Nesje, A., Kvamme, M. 1991. Holocene glacier and climate variations in western Norway: evidence for early Holocene glacier demise and multiple Neoglacial events. *Geology* 19: 610–12.

37 Berger 1978.

38 Fairbridge, R.W. 1987. Little Ice Age. In *The Encyclopedia of Climatology,* ed. J.E. Oliver and R.W. Fairbridge. New York: Van Nostrand Reinhold, 547–51.

39 Yoshino, M.M., Xie, S. 1983. A preliminary study on climatic anomalies in East Asia and sea surface temperatures in the North Pacific. *Tsukuba University Institute of Geoscience Annual Report* A4: 1–23.

40 Lamb, H.H. 1984. Some studies of the Little Ice Age of recent centuries and its great storms. In *Climatic Change on a Yearly and Millennial Basis,* ed. N.A. Mörner and W. Karlén. Dordrecht, Holland: D. Reidel, 309–29; Weikinn, C. 1965. Katastrophale Dürrejahre während des Zeitraums 1500–1850. *Acta Hydrophysica (Berlin)* 10: 33–54; Pettersson, O. 1912. The connection between hydrographical and meteorological phenomena. *Royal Meteorological Society Quarterly Journal* 38: 173–91; Lindgren, S., Neumann, J. 1981. The cold and wet year 1695: a contemporary German account. *Climatic Change* 3: 173–87.

41 Ludlum, D. 1966. *Early American Winters: 1604–1820*. Boston, Mass.: American Meteorological Society: Baron, W.R. 1982. The reconstruction

of eighteenth century temperature records through the use of content analysis. *Climatic Change* 4: 385–9; Catchpole, A.J.W., Ball, T.F. 1981. Analysis of historical evidence of climatic change in western and northern Canada. In *Syllogeus 33: Climatic Change in Canada*. Ottawa: National Museum of Natural Science, 96–148.

42 Wang, S.C., Zhao, Z.C. 1981. Droughts and floods in China 1470–1979. In *Climate and History: Studies of Past Climates and Their Impact on Man*, ed. T.M.L. Wigley, M.J. Ingram, and G. Farmer. Cambridge University Press, 271–88.

43 Yamamoto, T. 1972. On the nature of the climatic change in Japan since the "Little Ice Age" around 1800 AD. *Japanese Prog. Climatology* 97–110.

44 LaMarche, V.C., Jr. 1975. Potential of tree rings for reconstruction of past climatic variations in the Southern Hemisphere. In *Proceedings of WMO/IMAP Symposium on Long-Term Climatic Fluctuations*. Geneva: World Meteorological Organization, 21–30.

45 Porter, S.C. 1975. Equilibrium-line altitudes of late Quaternary glaciers in the Southern Alps, New Zealand. *Quaternary Research* 5: 27–47; Porter, S.C. 1981. Glaciological evidence of Holocene climatic change. In *Climate and History: Studies of Past Climates and Their Impact on Man*, ed. T.M.L. Wigley, M.J. Ingram, and G. Farmer. Cambridge University Press, 82–110; Hastenrath, S. 1981. *The Glaciation of the Equadorian Andes*. Rotterdam: Balhema; Broecker and Denton 1989.

46 Broecker and Denton 1989.

47 Jones, P.D., Wigley, T.M.L., Wright, P.B. 1986. Global temperature variations between 1861 and 1984. *Nature* 332: 430–4.

48 Wigley, T.M.L., Angell, J.K., Jones, P.D. 1985. Analysis of the temperature record. In *Detecting the Climatic Effects of Increasing Carbon Dioxide*, ed. M.C. MacCranken and F.M. Luther. Washington, DC: U.S. Dept of Energy, 55–90; Hansen, J., Johnson, D., Lacis, A., Lebedeff, S., Lee, P., Rind, D., Russell, G. 1981. Climate impact of increasing atmospheric carbon dioxide. *Science* 213: 957–66.

49 Charlson, R.J., Lovelock, J.E., Andreae, M.O., Warren, S.G. 1987. Oceanic phytoplankton, atmospheric sulfur, cloud albedo and climate. *Nature* 326: 655–61.

50 Schwartz, S.E. 1988. Are global cloud albedo and climate controlled by marine phytoplankton? *Nature* 336: 441–4; Wigley, T.M.L. 1989. Possible climate change due to SO_2-derived cloud condensation nuclei. *Nature* 339: 365–7.

51 Karl, T.R., Baldwin, R.G., Burgin, M.G. 1988. *Time Series of Regional Seasonal Averages of Maximum, Minimum, and Average Temperature, and Diurnal Temperature Range Across the United States 1901–1984*. Historical Climatology Series 4–5. Asheville, N.C.: National Climatic Data Center.

52 Henderson-Sellers, A. 1989. North American total cloud amount variations this century. *Palaeogeography, Palaeoclimatology, Palaeoecology* 75: 175–94.
53 Berner, R.A., Lasaga, A.C., Garrels, R.M. 1983. The carbonate–silicate geochemical cycle and its effect on atmospheric carbon dioxide over the past 100 million years. *American Journal of Science* 283: 641–83.
54 Stone, P.H., Risbey, J.S. 1990. On the limitations of general circulation models. *Geophysical Research Letters* 17: 2173–76.
55 Lindzen, R.S. 1990. Some coolness concerning global warming. *Bulletin of the American Meteorological Society* 71: 288–99.

4 World Hunger, Livelihoods, and the Environment

GITA SEN

INTRODUCTION

Environmental problems can probably be tackled most easily with regard to both analysis and policy when they can be penned within the purview of one or another of the natural sciences. Although the set of such pure (and simple) cases may be null, there is certainly a continuum ranging from problems that can be handled through primarily technical solutions to those in which social, economic, political, and technical aspects are so interwoven that a more holistic (and inevitably messier) approach is unavoidable. Until the 1960s, the problem of world hunger and food availability was viewed in the main as a technical problem – one, moreover, that appeared to be on the verge of a lasting solution based on the performance and promise of high yielding varieties of seeds and the associated "green revolution."

The problems of food availability are, however, clearly still with us in the last decade of the century, and have, if anything, become more acute in some parts of the world. Some would argue that the problem is really not one of per capita food production at a global level; rather, the problem is one of inadequate and distorted distribution in a world whose grain markets operate under oligopolistic conditions that are not conducive to rapid or efficient movements of grain from surplus to deficit regions in times of need. Lappé and Collins,[1] Morgan,[2] George,[3] and Bernstein et al.[4] are all critics of the food aid policies of major bilateral donors, which they feel have reduced self-provisioning in recipient countries and increased the unreliability of food supplies to the poor. The distributional problems inherent in existing market structures, poor physical infrastructure, and inappropriate social institutions and government policies are viewed as exacerbating the negative effects of unequal and skewed landholding patterns. Powerful as these arguments are, they sometimes ignore the linkages between environmental and public health concerns and nutritional systems; these are the focus of this chapter.

A HISTORICAL LOOK AT FOOD SUPPLY AND DEMAND

Much of the discussion of current hunger in the poorer regions of the world is couched in the Malthusian terms of population growth rates outstripping the growth of food supply. Since rapid population growth in these regions is itself a relatively recent phenomenon caused by sharp falls in mortality rates during this century, the problem of chronic hunger (as opposed to recurrent famines) has come to be popularly viewed as largely a problem of the poorer countries during the latter half of this century. An examination of the modern history of Europe contradicts this belief. There was considerable chronic hunger in many parts of Europe well into the early part of this century. More importantly, perhaps, malnutrition persisted among substantial sections of people in a number of European countries long past the time when national food supplies became adequate in the aggregate.[5]

We have little data from which to infer national food availability or consumption levels prior to the nineteenth century. Recent historical research has, however, made it possible to conclude with some degree of confidence that, as late as the early nineteenth century, daily food supplies in terms of calories, total protein, and animal protein in countries such as Norway, France and Germany were below currently accepted nutritional norms. (There is now considerable questioning of existing nutritional norms, especially with regard to proteins, particularly animal proteins. I do not enter into this controversy here, although it would obviously have some bearing on our reading of the historical and contemporary evidence.) This picture changed considerably during the next few decades. By the middle of the century, both calorie and protein availability had improved quite remarkably and diets had begun to change as well.

Nevertheless, malnutrition persisted in large sections of the population as a consequence of inequalities in income distribution and very low wage levels for the poorest sectors. In England the well-known reports of school inspectors, as well as the failure of would-be military recruits to meet physical standards, testified to the persistence of malnutrition and undernutrition well into the early part of the twentieth century. A League of Nations report pointed to the prevalence of scurvy, rickets, and anaemia all over Europe in 1936. Considerable malnutrition was also noted by government reports during the same period in the United States.

According to Grigg, an acknowledged authority on historical demography,

> it was not until the period since the end of the Second World War that malnutrition declined to a very small proportion of the population of the developed world. The principal reason for the earlier widespread malnutrition was poverty; substantial proportions of the population had incomes too small to buy an adequate diet ... Thus it is clear that the provision of adequate national food supplies is not sufficient to ensure the elimination of undernutrition and malnutrition. The poor must have enough money to purchase the minimum diet. It is disturbing to note that there was a lag of up to a century between some European countries' achievement of an adequate national food supply and elimination of malnutrition.[6]

Estimating nutritional need and comparing it with food availability for developing countries in the post–Second World War period is made difficult both by conceptual problems and by the inadequacy of data. Daily caloric requirements clearly vary in accordance with body size and biological or work demands. However, even norms by age and gender have been challenged on the grounds that the range and spread of metabolic rates among individuals make such norms highly suspect.[7] Such challenges do not, in my opinion, overturn the presumption of considerable malnutrition, especially in the low-income countries. They do, however, indicate the need for caution in interpreting specific estimates of nutritional status. Compounding this problem is the fact that the quality of data on the production and marketing of food varies widely across countries; these data are subject to considerable error even in the better-organized statistical systems. For instance, nutritional data for the state of Kerala in India have ignored the importance of tapioca in the average local diet, thereby seriously underrepresenting the nutritional level in the state. In a number of African countries, the presence of governmental intervention in food grain markets can lead to underestimates of production, particularly when output is estimated on the basis of the throughput of government controlled markets, while there is a significant unrecorded parallel market in grain.[8]

Table 4.1
Rate of Change in Food Output Per Capita (per cent per year)

	1952–54 to 1959–61	*1961–70*	*1971–80*
Africa	–0.2	0.1	–1.2
Far East	1.1	0.9	0.9
Latin America	0.3	0.8	1.2
Near East	0.8	0.3	0.6
Asian CPE[1]	No data	0.9	1.6
All developing	0.7	0.7	1.0
All developed	1.7	1.4	1.1
World	1.1	0.8	0.6

Source: Grigg 1935, Table 5.5
[1] Centrally planned economies

Despite the possibility of underestimates on the production side, comparisons of broad regional data on population growth and food output for the period 1950–80 indicate that the latter kept pace with the former for the developing countries taken as a whole. Africa was the only major region for which this generalization did not hold. Although there was considerable intraregional variation in the growth of per capita food output, any simple notion of population growth outstripping food supplies is not borne out by the evidence. More complex factors linked to asset and income distributions, market structure and food policies appear to be major determinants of food access and nutritional status (see Table 4.1).

During the 1980s, a decade of economic crisis in much of Africa and Latin America, food production per capita dropped in a number of countries. The average index of food production per capita fell in half of the forty-two low income countries during the period 1979–81 and 1986–88.[9] This, at least in part, is a reflection of the severe financial crisis besetting many of the same countries; consequently, the solution to the food problem cannot be independent of more general economic policies.[10]

THE SUPPLY SIDE

Much development thinking since the 1960s has been dominated by a concern to increase food supplies. One result of this has been the well-known "green revolution" in major crops such as wheat and rice. The technological promise of combinations of high-yielding hybrid seeds, chemical fertilizers, pesticides, and improved water supply was brought to bear on agricultural production in a large number of countries, especially in Latin America and Asia. In both regions output increases have come from increases in crop area as well as yield, but the relative importance of the two factors has been different.

One major reason for this was the difference in landholding patterns carried over from the colonial period. In Latin America landholding was generally much more unequal than in Asia. Prior to the 1960s, around two-thirds of farm area in Latin America was in estates of over 1000 hectares in size, large parts of which lay idle or in fallow. Monoculture was common, livestock rearing was rarely combined with crop culture, and even crop rotation or the use of green manure was rare. Most of the farm population, on the other hand, was crowded onto very small and unviable tenancies on marginal lands, producing part of their subsistence, and providing labour to the big estates.

The first half of this century saw considerable extension of the frontier with the expansion of export crops into new areas, the growth of transport, and the control of endemic diseases such as malaria. But major increases in cultivable area came after 1950, partly in response to growing pressure for land reform. Between 1950 and 1980, arable land area increased from 86 million to 162 million hectares, and the area in major food crops doubled. Reduction of fallow and an increase in multiple cropping have been relatively unimportant in this area expansion, and yield increases accounted for only 40 per cent of the increase in food output in the 1970s.[11]

In Asia, most of the post-1960 increase in food output came from increased yield or increases in gross cropped area through multiple cropping. Extending the margin of cultivation was important only in a few Southeast Asian countries such as Indonesia, Thailand and Malaysia. During the 1970s over 70 per cent of the increase in cereal output in south and southeast Asia came from higher yields.[12] Average crop yields in Asia (excluding China and Japan) increased by 54 per cent in wheat, 43 per cent in rice, 39 per cent in maize, and 30 per cent in millet and sorghum between 1950–59 and 1970–79.[13]

These increases in the recent past in Latin America and Asia have clearly made it possible for overall food output to keep pace with the growth of population. What have been the costs, and what are the distributional implications? Although there has been considerable debate on the distributional aspects of the new technologies, there has been little attempt to assess their social costs relative to the benefits of increased output. Part of the problem arises from the fact that some major costs have been assumed in the past and their present value is difficult to quantify.

In many countries the technology has been concentrated in the regions already best supplied with a major input such as water. In

India, for instance, the choice of the northwest (Punjab, Haryana, and west Uttar Pradesh) was dictated in part by the availability of water. However, it was the major investment in irrigation by the colonial state at the turn of the century that gave the previously dry zone of the Punjab its later advantage. Obviously, such historical costs cannot be accounted for in an assessment of the social returns from the new technology; however, any extension of the technology into currently dry areas will require major new investments unless technological breakthroughs in high-yielding dry crops can be made. Indeed, the narrow regional base of recent food increases in India becomes clear from the observation that the northwest accounts for about one-third of total grain production, almost all of the government's wheat procurement, and about two-thirds of rice procurement.

The true costs of the new technology are also often disguised by governmental interventions in the markets for both inputs and outputs. Subsidized inputs of fertilizer, pesticides, seeds, water, and power increase the private profitability of the new crop varieties, while governments also set floor or ceiling prices for grain. These interventions do not always work in favour of farmers; in India, for example, the terms of trade between foodgrains and manufactured goods have tended to turn in favour of the latter after the mid-1970s. But what is important for this discussion is that it is extremely difficult to assess the true social costs and benefits of the new technology.

An additional problem resulting from such a high incidence of subsidies is the dependence of food output on fiscal manipulations which leave growers open, especially in times of fiscal stringency, to the possibility of wide fluctuations and drops in output. This exacerbates the heightened variability in output arising from the greater requirements of water for the new crop varieties in regions which are heavily rain-dependent.

Greater output instability possibly worsens the distributional problems that critics associate with the green revolution technology. In itself, the technology has few economies of scale, since (in theory) it can be adopted equally by both small and large farms. However, its substitution across the board of purchased inputs for farm-produced ones handicaps smaller farmers, who cannot adopt this practice to the same extent as larger farmers. Moreover, once the larger farmers in a region begin to switch over to the high-yielding hybrid varieties, profits, land prices, and rents all start going up. Smaller and tenant farmers have little choice at this point but to switch over as well in the hope of making enough profits to stay abreast of increased costs. Such farmers may also have to pay differentially higher prices for inputs such as credit obtained through informal markets. These pressures on small farmers are heightened if the terms of trade turn against agriculture, and/or if governments are forced to reduce their subsidies on inputs.

Even more serious distributional questions have been raised in regard to the situation of landless agricultural labourers. Landless households account for over 25 per cent of agricultural households in South Asia; if we include households with very marginal amounts of landholding, this figure rises to over 50 per cent. The real earning of agricultural wage workers is a function of money wages, prices, and of the number of days of employment they can get in a year. Although the evidence varies across countries and regions, the overall generalization (except in regions of heavy mechanization) is that aggregate employment has increased while real wages have remained stagnant or gone down. The net impact is that, by and large, agricultural labourers have not shared significantly in the benefits of the growth generated by the new technology. In addition, growing commercialization tends to deny to the landless such traditional non-monetary perquisites as free fodder, fuel, use of waste lands, gleanings, etc. At very low levels of income this loss can mean a serious decline in consumption or an increase in the labour required to ensure the household's subsistence and reproduction.

Of particular importance for our discussion are the ecological questions that are increasingly being asked of the proponents of the green revolution technology. It is now acknowledged that, in much of Asia at least, traditional varieties and traditional methods, although incapable of producing the high yields of the hybrids, had a number of positive features. In the past, the long and slow processes of learning-by-doing led to crop and seed selections more appropriate to their agroclimatic environment. Since agriculture for the most part tended to be monsoon dependent, the ability of a crop variety to withstand drought or excessive rain was an important element in its selection. So also was its ability to ward off pests or disease and still provide a level of subsistence. The result was considerable biodiversity in agriculture; in India alone nearly 40,000 rice varieties have been grown.[14] Water storage and conservation were important, and sophisticated systems existed in many countries; irrigated agriculture, where it occurred, sustained rather than depleted the water table. Crop rotations, green manure, and, in some regions, the use of dung returned or fixed nutrients in the soil.

Unfortunately, the green revolution technology has run counter to many of these ecologically sound aspects of traditional agriculture. Its heavy use of water has led to problems of salinity in some regions, as well as to a general depletion of water tables through the tapping of ground water sources as a way of escaping the vagaries of the rains. Monocropping, insufficient crop rotation, and the excessive use of chemical fertilizer have reduced the fertility of soils, creating a vicious circle wherein the farmer has to use more and more fertilizer merely to maintain yields. Hybrid varieties do not have the resistance to pests and disease of the traditional varieties, and pesticides have been used and overused by farmers who are insufficiently aware of their dangers.

The pollution of falling water tables by both fertilizers and pesticides is a serious problem, especially in rural areas where people depend on the same groundwater sources for their daily needs. The health effects of indiscriminate chemical use on usually unprotected agricultural labourers are only now beginning to emerge through a few microstudies; typically, this potentially (and, in all likelihood, actually) major health problem has been disguised by the fact that many of the workers in green revolution areas tend to be seasonal migrants. Population flux and the weak capacity of health infrastructures have combined to submerge a problem of both general and reproductive health.

A final issue raised by the switch to hybrid varieties is the loss of biodiversity. This issue has typically been discussed in the context of the destruction of tropical rainforests, the repository of much of the planet's genetic variety. The transformation of tropical agriculture is a less widely recognized but nonetheless significant contributor to the same problem. Compounding the loss of variety is the loss of control and knowledge by local farmers, who now must buy seed each season rather than setting aside part of the harvest for the purpose. In a world without political economy this would perhaps not be a major problem, since the gene pool could be preserved in frozen storage. In the real world there is legitimate concern about the consequences of such severe dependence on centralized (multinational corporate or national governmental) preservation on local farmers' already curtailed ability to exercise choice or judgment about ecologically or economically appropriate crop varieties.

Many of the problems associated with the green revolution are not specific only to tropical agriculture. Similar environmental questions have been raised in the context of North American agriculture, for example, but have been difficult to solve up to this point; both economics and the politics of government policy-making have so far been at cross purposes with ecological concerns. However, limiting our present discussion to the problems faced by developing countries, we must consider two additional factors that appear to necessitate the push for higher-yielding domestic agriculture, namely, population growth and the politics of food aid. For instance, both India in the mid-1960s and Bangladesh in 1974 experienced considerable pressure from their main food aid donors, in the face of severe droughts and food-stock depletion, to bring about major changes in economic policy and political stance in return for food supplies. Both had to give in, but India went on with stronger determination to adopt the new technology so as to become nationally self-reliant in foodgrains by the 1980s. (The concept of food self-sufficiency is controversial. People may still go hungry because of poor distribution of wealth, incomes, and food supplies even when there is an adequate total supply.[15]) In Bangladesh the fledgling social democratic government fell, and subsequent governments have been unable to reduce the country's vulnerability to food aid pressures.

Population growth is unlikely to be reduced by increased food production, but there can be little doubt that it is the increased output resulting from area increase and higher yields that has made it possible for Asian and Latin American agriculture to keep pace with population growth in the aggregate. However, the distributional and environmental issues raised earlier suggest the need to develop a concept of sustainability and sustainable development that incorporates not only technological concerns but also economic and political ones.

The supply problem in Africa is rather different from that in both Asia and Latin America. The ecological base in Africa for increased cultivated area or yield has traditionally been perceived to be fragile. Geological stability and the low level of volcanic activity made soils thinner and more subject to leaching. The agricultural response to this and other problems (such as the difficulty of maintaining cattle for manure in tsetse-infested regions) was shifting cultivation with long fallow and intercropping, which provided crop variety, staggered harvest times, and lowered soil temperatures through the use of shrubs to protect lower-growing crops. There has been considerable discussion in the literature on African agriculture about the shift to shorter (bush) fallow systems and its link to population growth, labour use, and productivity.[16]

One of the more innovative approaches to the technological complexity of African agriculture is the "farming systems" approach, which has become increasingly popular among aid agencies during the last decade.[17] This approach may be viewed as a way of dealing with the problem of appropriate technology by adapting research and extension systems to the microenvironmental and technological needs of farmers through greater emphasis on feedback from the ground up. Its proponents are also more sensitive to the micro organization of labour inputs, particularly with regard to gender. While this represents a considerable improvement over earlier approaches, it still gives primary emphasis to technological solutions.

But the problem of increasing agricultural output in Africa is not simply one of fragile soils unable to cope with rising population density, as it has often been portrayed. Land use patterns, systems of fallow, crop choices, and land tenure all underwent major changes during the colonial era, many of which have not been reversed. These changes shifted land use and crop composition towards export crops, reduced the availability of labour and other inputs for subsistence food crops, and fundamentally altered the ecological balance of agricultural production. The declining viability of the continent's agriculture in the 1970s and 1980s was not a simple matter of population outstripping the growth of food supply; export production also became less viable during a period when the terms of trade turned against primary exporters. This crisis in external viability translated into fiscal pressures on governments, who were then coerced into structural adjustment programs in exchange for balance-of-payment support. This is not to deny the weakness of governments or the inefficiency of parastatals, but simply to provide some balance to our perception of Africa's food problem.[18]

The highly influential Berg Report[19] made the argument that African agriculturists should specialize according to their historically developed comparative advantage. Focusing on export agriculture and removing government interventions in markets were proposed as mechanisms to free the productivity of farmers. Unfortunately, the language of Reagan-era free markets accorded poorly with the realities, ecological and economic, of farmers on the ground, as has now been noted by many commentators.[20] The UN Economic Commission for Africa has argued for the attainment of a level of food self-sufficiency as a means of tackling the severe crisis of basic needs in the region and of establishing a basis for more viable economic strategies in the long term. In a way, the approach the Commission suggests is similar to that typically adopted by traditional farmers in drought-prone regions. It is based on the realization that in an economic/ecological situation that is both highly uncertain and highly risky, only the well-resourced or the foolhardy can afford to opt against self-reliance. Long term sustainability means conservative self-provisioning.

In its support for rebuilding European agriculture during the Marshall Plan years, U.S. foreign policy recognized the validity of food self-sufficiency as an important basis of strong economies, even though this went against the commercial interests of the American farm lobby at the time. But Europe was then a terrain for intense political contestation with the Soviet Union, and foreign policy interests were allowed to override narrow commercial interests. Unfortunately, there is no comparable situation currently at work in favour of African self-provisioning.

I have tried to argue that the problem of food supply is not susceptible to simple technological solutions, nor is it only a matter of population outstripping production. A reexamination of the nineteenth-century public debate over the English Corn Laws would prove the point. Complex questions of land tenure, labour use, and government policies towards inputs, outputs, and technology, as well as towards prices and exchange rates, determine the conditions of farm viability. On the other hand, the demand for food is beset by equally complex issues, some of which I will now describe.

THE PROBLEM OF ACCESS

Food provisioning is an area in which the economist's basic tool – effective demand – fails miserably to do the job. If our concern is with meeting reasonable nutritional standards for all, then clearly the problem of access to food is not adequately captured by the concept of demand made effective by purchasing power. Indeed, the food problem could be seen on one level as precisely the problem of those who are too poor to translate their need for food into market demand.

The question of access has been most persuasively treated through Sen's concept of "entitlements."[21] Through a historical examination of older and more recent famines, Sen shows that insufficient food availability in the aggregate is almost never the cause of famine mortality. Rather, social and economic entitlements that guarantee access to food break down for a variety of reasons. Some vulnerabilities are traditional and culturally sanctioned, such as those that allot a lesser quantity and poorer quality of food to girl children and women in many parts of South Asia. Others are political, as in the colonial government's cornering of rice stocks during the Bengal famine of 1943. Still others, such as the Wollo famine in Ethiopia in 1974, are the result of the normal working of the market, whereby commercialization increases the vulnerability of poor people and regions without providing them with adequate buffers to withstand the increased risk.

Entitlements in this sense are clearly linked to class, gender, and other hierarchies which order access to resources, incomes, and consumption. In situations where markets do not function to allocate adequate food entitlements, public distribution systems become important. The entitlement approach to access is valuable because, *inter alia*, it highlights the policy implications of market inadequacies and brings the gender basis of food distribution within households into sharp relief. Sen's argument[22] is not about market failure, but about the ways in which perfectly functioning markets fail to meet basic needs and even increase vulnerability.[23]

It is also possible to argue from the above that different strategies for increasing food supply are not neutral in their effects on entitlements. An individual's food entitlement is likely to be best served when she or he has a direct and unmediated source of adequate income. This is particularly relevant to arguments in favour of women's economic independence. But the entitlements of all the poor in regions or countries are affected by both the distributional and environmental consequences of particular strategies towards supply. Thus, as argued earlier, green revolution technology may increase the vulnerability of the poorest; it certainly reduces their traditional entitlements of food, fuel, fodder, and water.

The entitlement approach to the question of access takes us a considerable way towards conceptually integrating food supply questions, environmental and public health concerns, and issues of gender and class. But gender is not only a matter for concern from the side of access; it also enters into the discussion of production. In many parts of sub-Saharan Africa women are farmers in their own right; although their direct access to land and labour is limited and has been eroded over time, they still have considerable scope for decision-making and contribute increasing amounts of labour.[24] Even in parts of the world where women are not independent farmers, they participate widely as agricultural labourers, but often at lower wages and with higher rates of unemployment and underemployment than men. The impact of agricultural strategies on wage labourers holds *a fortiori* for the women among them, who are usually the most vulnerable. There are now many examples of agricultural mechanization privileging men and displacing women.

Women are also especially susceptible to the negative environmental effects of production technologies because of their tasks in biological reproduction and household subsistence. For instance, the effects on the reproductive health of women labourers in areas of heavy pesticide and fertilizer use are not known. Similarly, the lowering and pollution of groundwater tables and the loss of entitlements to wasteland resources of food, fuel, and fodder makes women's labour in provisioning for these longer and more arduous.

Linking all these aspects of the food problem is a major challenge. However thinking about nutrition systems rather than food supplies

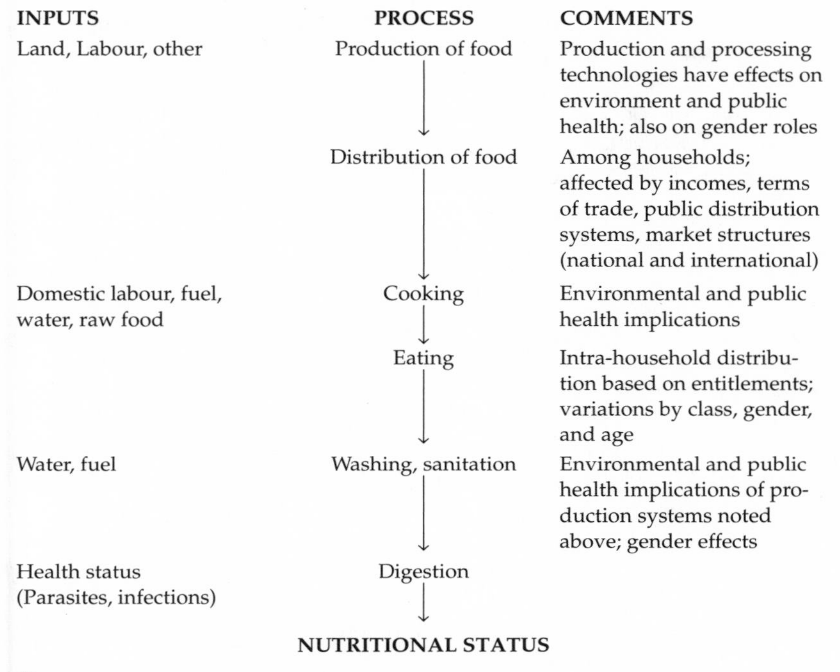

Figure 4.1
A diagrammatic representation of a nutrition system.

allows us to integrate production processes, questions of access, environmental and public health concerns, and gender hierarchies, and then to draw out their implications for nutritional status. A nutrition system involves looking at the full set of linkages from the level of aggregate food production to individual nutritional status. Working backward from the proximate determinants of an individual's nutritional status, one sees that environmental factors working through public health conditions, chemical and other pollution, water and fuel availability, and reproductive health appear to have effects at various levels of the nutrition system, starting from the level of production itself (Figure 4.1). Improvements in the nutrition standards of a population are predicated not only on food technologies and agricultural pricing policies, but on the gender, environment, and health implications of those same technologies and policies. A growing number of micro-level food production experiments are attempting in effect to use a nutrition system approach. For example, the Proshika Centre for Human Development in Bangladesh has been promoting organic farming through over seven hundred women's groups; field workers help villagers to develop more sustainable farming practices which are ecologically more conservative and which also focus directly on improving nutritional levels.[25] However, governmental food policies are often at cross purposes with such experiments; further progress depends on the extent to which the former can be made consistent with the promise of the latter.

NOTES

1 Lappé, F.M., Collins, J. 1988. *World Hunger: Twelve Myths*. London: Earthscan.

2 Morgan, D. 1979. *Merchants of Grain*. London: Weidenfeld and Nicolson.

3 George, S. 1977. *How the Other Half Dies: The Real Reasons for World Hunger*. Harmondsworth: Penguin.

4 Bernstein, H., Crow, B., Mackintosh, M., Martin, C. 1990 *The Food Question: Profits versus People?* London: Earthscan.

5 Grigg, D. 1985. *The World Food Problem* 1950–1980. Oxford: Basil Blackwell.

6 Grigg 1985, 52.

7 Sukhatme, P.V., Margen, S. 1982. Autoregulatory homeostatic nature of energy balance. *American Journal of Clinical Nutrition* 35: 355–65.

8 Centre for Development Studies 1975. *Poverty, Unemployment and Development Policy: The Case of Kerala*. New York: United Nations; Mbilinyi, M. 1990. Structural adjustment, agribusiness and rural women in Tanzania. In Bernstein et al. 1990.

9 World Bank 1990. *The World Development Report 1990*, Oxford University Press.

10 UN Economic Commission for Africa 1989. *African Alternative Framework to Structural Adjustment Programmes for Socioeconomic Recovery and Transformation*. (E/ECA/CM.15/6/Rev.3).

11 Grigg 1985, 190.

12 Ibid.

13 Ibid.

14 Farmer, B.H. 1978. The green revolution in South Asian ricefields: environment and production. *Journal of Development Studies* 15: 304–19.

15 Patnaik, U. 1990. Some economic and political consequences of the green revolution in India. In Bernstein et al. 1990.

16 Boserup, E. 1965. *The Conditions of Agricultural Growth*. Chicago: Aldine.

17 Fresco, L.O., Poats, S.V. 1986. Farming systems research and extension: an approach to solving food problems in Africa. In *Food in Sub-Saharan Africa*, ed. A. Hansen and D.E. McMillan, Boulder, Colo.: Lynne Reinner

18 UN Economic Commission for Africa 1989.

19 World Bank 1981. *Accelerated Development in Sub-Saharan Africa: an Agenda for Action*. Washington, D.C.: The World Bank.

20 E.g. Mackintosh, M. 1990. Abstract markets and real needs. In Bernstein et al. 1990.

21 Sen, A.K. 1981. *Poverty and Famines*, Oxford University Press; see also Dreze, J., Sen, A.K. 1990. *Hunger and Public Action*. Oxford University Press.

22 Ibid.

23 See Mackintosh 1990.

24 Mbilinyi 1990; Whitehead, A. 1990. Food crisis and gender conflict in the African countryside. In Bernstein et al. 1990.

25 Rush, J. 1991. *The Last Tree – Reclaiming the Environment in Tropical Asia*. New York: The Asia Society.

5 Species Impoverishment

M. BROCK FENTON

INTRODUCTION

Some biologists, naturalists, and conservationists maintain that the diversity of life forms and species is one general indicator of planet Earth's state of health. The loss of species, or "species impoverishment," is said to be proceeding at an accelerating rate, a trend that bodes ill for our future. The situation is aggravated because the earth consists of interconnected ecosystems. Ironically, the best evidence of interconnections comes from DDT. Although this insecticide was used in just some locations, its presence in animals throughout the world reveals just how extensive the connections between the globe's ecosystems and their components are.[1]

Every year, organisms disappear from our planet. At one level their disappearance represents the deaths of individuals; at another level, the extinctions of whole populations. Sometimes the vanished populations are considered to have been species. Extinct populations and species may be well known – e.g., the Beothuks of Newfoundland, the passenger pigeon (*Ectopistes migratorius*), and the dodo (*Raphus cucullatus*) – or obscure – e.g. the eastern brown, or grizzly, bear (*Ursus arctos*) and the sea mink (*Mustela macrodon*). Extinctions also occur above the species level; dinosaurs and trilobites are fine examples of once successful groups of animals that disappeared leaving no descendants. The death of individuals and the extinction of species are part of the natural process of evolution[2] and virtually all of the plant and animal species that have ever existed may now be extinct.[3] Not surprisingly, some people use this fact to allay concerns about the disappearance of organisms. While I agree that humans are a "natural" part of the global system, we are beyond using a natural process argument to justify continued eradication of other organisms. The magnitude of human populations, their access to technology, and their selfish large-scale manipulations of systems take them out of the realm of the natural processes that have been part of the earth's history. The other chapters in this volume explore many aspects of these disruptions.

Saving species from extinction is a stated goal of many conservation organizations. The purpose of this chapter is to consider different aspects of species impoverishment. I will begin by reviewing some features of the nature of "species" and of extinction, and proceed to the glimmers of hope provided by evidence of the recuperative power of species and ecosystems. Then I will review some of the factors that threaten organisms' survival before addressing the role that conservation can play in reducing these threats. Using some examples, I will reflect on approaches that could offer solutions to some of the conservation problems confronting humanity; ironically, these approaches centre on the selfishness of human beings, arguably the main cause of species impoverishment today. While I have drawn my examples from terrestrial animals, usually birds and mammals, the same basic principles apply to all other living organisms.

THE SPECIES IN BIOLOGY

The species is a basic unit in biology, the identifier for any of the organisms formally known to science. Biologists depend upon the Linnaean system of nomenclature to provide a distinct name for each species that has been identified. Species are the immediate product of evolution, and so the biological species concept generates various disagreements.[2] Most first-year university biology students know that a species consists of groups of organisms that actually or potentially interbreed to produce fertile offspring. These groups or populations are reproductively isolated from other such groups. Reality, however, may be more complex, partly because hybrids sometimes form in areas where the ranges of two species meet.[4] Putting aside organisms that do not reproduce sexually and others with alternative (from a human perspective) reproductive lifestyles, this traditional definition fits comfortably in some cases and not in others.

For example, some species occur over relatively small areas, show little variation, and are not divided into subpopulations, such as the spotted bat (*Euderma maculatum*; Figure 5.1A).[5] More widespread

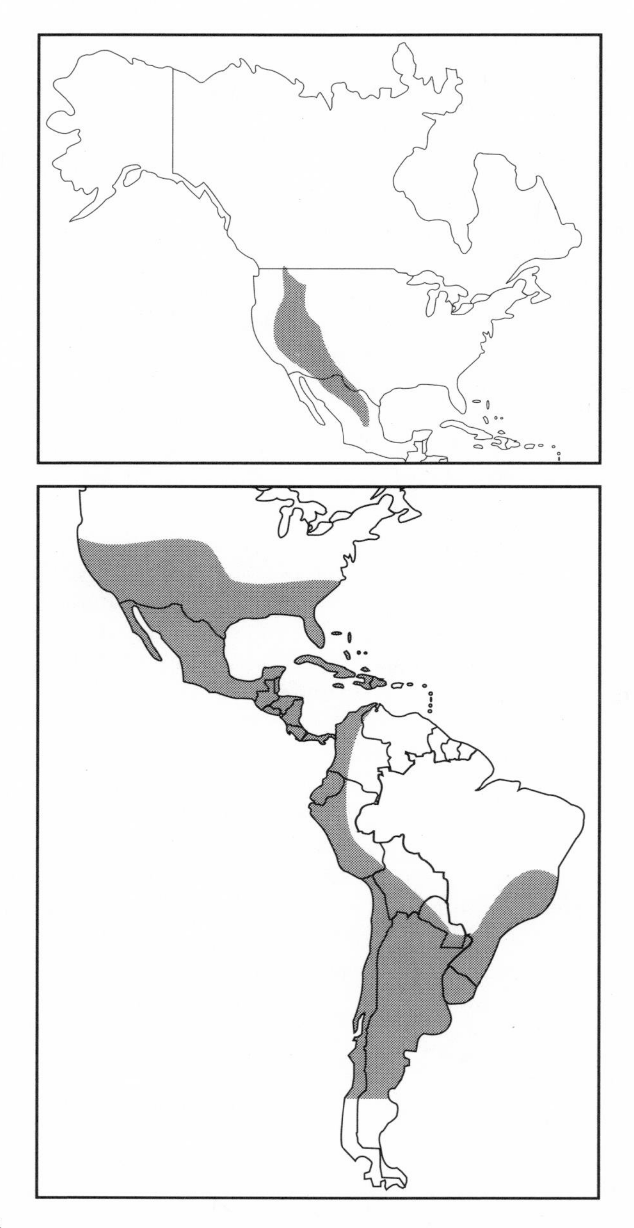

Figure 5.1
An example of species distribution and its effect on species variation. (A) The relatively limited range of the spotted bat. No subspecies are known. (After Van Zyll de Long 1985.) (B) The more extensive range of the Mexican free-tailed bat. Nine subspecies are known. (After Wilkins 1989.)

species often show more variability and are divided into subspecies. For example, Mexican free-tailed bats (*Tadarida brasiliensis*; Figure 5.1B) are grouped into nine subspecies,[6] reflecting both their variation and their extensive geographic range. The species concept imposes an artificial classification on systems with natural variability. Two practical questions are (1) "At what level can a population be identified as a subspecies?" and (2) "When do you draw the line between subspecies and species?" The answers are not straightforward, for among biologists there is general agreement about the overview, but not about details.

Domestic dogs illustrate the problem. All dogs belong to one species, *Canis familiaris*. Thus, although dogs show tremendous variation in size, shape, colour, and behaviour, all of them should have the potential to interbreed and produce fertile offspring. Is this always possible? A male hairless chihuahua would need a stepladder and some help to achieve this with a female Newfoundland.

But why does this matter to anyone except biologists working with the organisms in question? There are two important reasons. First, names give organisms identities and, for all practical purposes, nameless organisms do not exist. An unfortunate consequence of this is that any organisms not formally described by biologists have no name and no status with respect to conservation. Second, in the United States of America, the Endangered Species Act of 1973 protects plant and animal species at three levels: species, subspecies, and populations. This act offers no protection to unrecognized populations. The Convention for International Trade in Endangered Species (CITES) is intended to protect populations by stopping trade in described species that are identified as endangered. Since some biologists estimate that at least one million species remain to be "discovered," many organisms cannot be protected.

The names of species can generate other problems with respect to legislation and legal status.[7] One example is provided by the Florida panther or cougar (*Felis concolor coryi*). About fifty of these animals remain in southern Florida, apparently showing significant physiological and reproductive effects of inbreeding depression. Between 1957 and 1967, the release of some animals from captivity into the Everglades appears to have introduced genes from South American *Felis concolor* stock into the Florida population.[8] The prevailing and precedent-setting interpretation contends that the Endangered Species Act does not protect hybrids. So Florida panthers, now genetically contaminated, are no longer protected by the Endangered Species Act.[9]

This policy concerning hybrids has already affected the status of some other animals and could threaten their survival in the longer term. For example, gray wolves (*Canis lupus*) and red wolves (*Canis rufus*) are known to be contaminated with coyote (*Canis latrans*) genes. Dusky seaside sparrows (*Ammodramus maritimus nigrescens*) were also genetically contaminated and with the removal of protection have already disappeared.[10] The status of red wolves as a distinct species is questionable, a problem that is directly affecting its future.[11]

EVIDENCE OF RECUPERATION

Living systems show remarkable powers of recuperation. By 1930, beavers (*Castor canadensis*) were all but exterminated over much of their range in North America. Today their populations have recovered dramatically and in many parts of their range beavers have become nuisances. By 1940 there were fewer than 100 white rhinos (*Ceratotherium simum*) in all of southern Africa, but by 1989 their numbers had surpassed 4000. Setting aside, for the moment, the genetic consequences of such a recovery, large and small mammals can reestablish their populations when given adequate protection.

Recuperation, however, extends beyond individual species. By 1880, approximately 80 per cent of the land area of New York State had been deforested. Within one hundred years the forests had rebounded to cover 80 per cent of the state's land area.[12] While it is obvious that the New York State forests of today probably differ from those of two hundred years ago, the changes to the landscape are heartening.

There is similar evidence from elsewhere. For example, when J.L. Stephens and F. Catherwood visited the Yucatan Peninsula in 1820 they found extensive Mayan ruins, some overgrown by jungle, others by scrub vegetation. Some of these sites had been abandoned less than two hundred years before and had been overgrown in the intervening period. Today, tourist developments have brought repeated deforestation to many of these sites. Hammond[13] points out that large areas of lowland rain forest in the territory of the Mayans had probably been cleared at the height of that civilization. This means that the recovery of the disturbed systems had been very extensive indeed.

There are similar records of ecosystem recovery around other human settlements. For example, the civilization that produced Great Zimbabwe in southern Africa was in decline by 1600 and its ruins had been completely overgrown by 1850.[14] Today, as in the case of the Mayan ruins, development of the site has again removed the forest.

While the recuperative powers of species and systems provide hope for the future, they cannot overcome extinction. Nothing brings back extinct species.

VULNERABILITY

A general equation can be used to calculate the rate at which a population will increase in numbers over time:[15]

$$\frac{dN}{dt} = rN\left(\frac{K-N}{K}\right).$$

This equation also identifies some important areas of species vulnerability. N represents the number of individuals in the population. The expression (dN/dt) represents the change in N over a short time period, t. The intrinsic rate of increase, r, represents how quickly a species can reproduce under ideal circumstances. K represents the carrying capacity of the environment, i.e. the maximum number of individuals it can support.

The calculation of r takes into account the incidence of reproducing individuals in the population as well as fertility, fecundity, natality, mortality, age at first reproduction, survival of individuals, and so on. The intrinsic rate of increase is the difference between recruitment into and departure from the population. All other things being equal, a species that produces a litter of four young three times a year, for example, will have a higher r value than a species producing one litter of four each year. Detailed knowledge about the biology of a species will be essential for the accurate estimation of r. Unless something changes, species are doomed to extinction when the rate at which individuals are disappearing exceeds their r value.

Decisions about a species' well-being depend upon knowing N. In the case of our own species, however, accurate information about population size is elusive, even in industrialized countries. The problem is exacerbated when the populations in question are rare and elusive, as Sudman et al. noted for some groups in American societies;[16] these problems are much amplified when other species are the subject of the census. Density is another way to express population. At low population densities, some species will not achieve their potential r values because of the logistical difficulty of finding mates. At very high population densities some organisms will not achieve maximum r values because social pressures may depress achieved r.[17]

The carrying capacity, K, is an easy concept to understand: it is simply the capacity of a system to support a population of organisms.

Putting a value on it is another matter entirely. Carrying capacity may be determined by the availability of food or of places to live, or by many other factors according to the species and the situation. Accurately estimating K for any species requires detailed knowledge of the biology of the organism in question and the ecosystem in which it occurs.

Unlike virtually all other organisms, humans have disrupted the natural relationship between the sizes of their populations and the biological carrying capacity of the areas where they live. This disruption occurs when humans import biological resources such as food and generate pollution by releasing the waste of imported food into ecosystems that did not produce it.

From here we can see how the different direct causes of species impoverishment relate back to matters of r, N, or K.

Outright Destruction

Humans have a long record of hunting species to extinction. Great auks (*Pinguinis impennis*), Steller's sea cows (*Hydrodamalis gigas*), dodos, and passenger pigeons are prime examples of species extirpated by human harvesting. The process continues today.

Large animals, particularly mammals that may eat humans or damage their property or crops, are especially vulnerable.[18] One reason for their vulnerability may be that these animals have always occurred in relatively small numbers. In the last four hundred years the North American ranges of grey wolves (*Canis lupus*) and wolverines (*Gulo gulo*) have shrunk dramatically in the face of increasing human pressure. Some of this decline in numbers arises from conflict with human activities. Brown (grizzly) bears have faced the same reduction in numbers and each year there is some public outcry when tourists are killed or maimed by wild bears.[19]

Sometimes market demand for an animal product generates harvesting rates that exceed r. The black rhinoceros (*Diceros bicornis*) is a current example. It is estimated that in 1960 there were 60,000 black rhinos in the wild in Africa (Figure 5.2) and the species was one of the "big five" on the list of "Great White Hunters" from abroad. By 1990, there were fewer than 5,000 black rhinos, most of them in the Zambezi Valley in Zimbabwe. Rhino horn, whether from black rhinos or from other species, has long been prized as material for making dagger handles in North Yemen. In India and the Far East there is a demand for powdered rhino horn as an ingredient in fever medications and aphrodisiacs. Black rhinos are being killed for their horns and for no other reason. Although in Zimbabwe teams of government

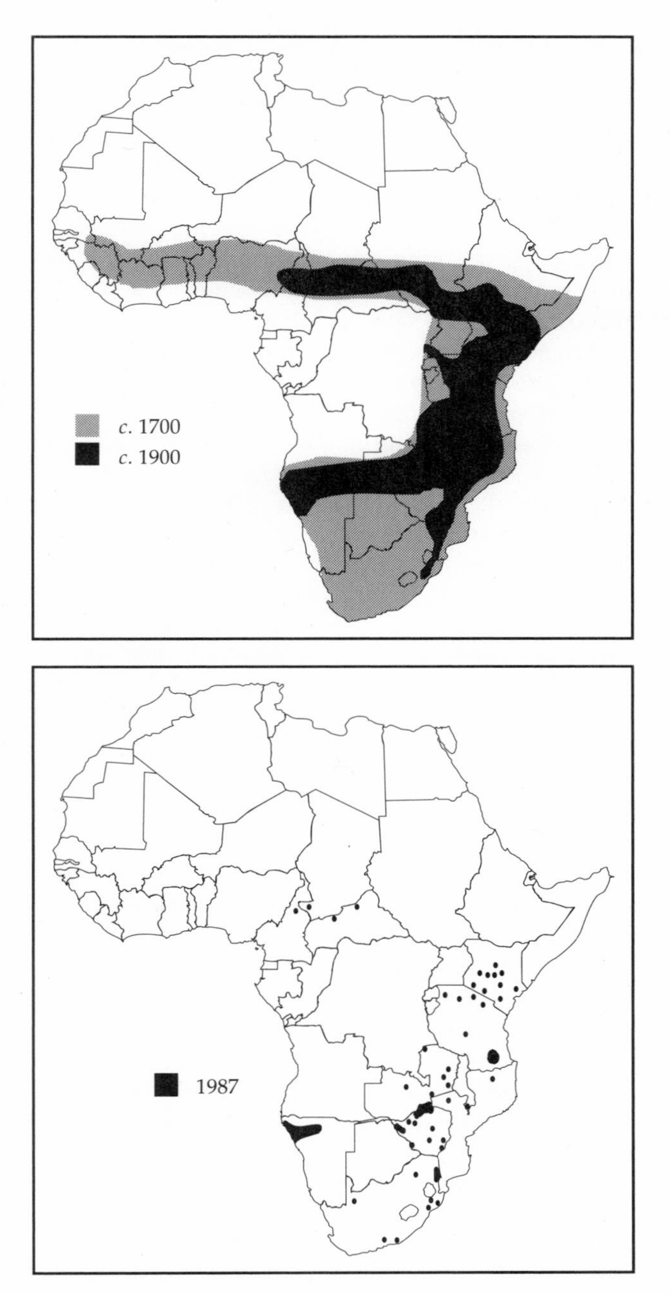

Figure 5.2
The geographic range of the black rhinoceros, *c.* 1700, *c.* 1900, and in 1987, showing progressive reduction in distribution. (After Cumming et al. 1990.)

parks and wildlife personnel actively hunt rhino poachers, whom they are authorized to shoot on sight, the killing of rhinos continues. In other parts of the world, other species of rhinoceros face similar lethal pressure generated by supply-and-demand economics.[20] In some areas, wild rhinos have been dehorned to save them from poachers. This move presumes that the horns are not vital to the rhinos' survival (the horns may have a role in defence, whether against predators or competitors).

The situation raises several important questions. Will any rhinos survive to the turn of the century? If so, will they be creatures of zoological gardens in areas where they can be protected like crown jewels or other precious commodities? How many rhinos constitute a viable population? Are rhinos worth saving?

Humans can cause extinctions by introducing alien organisms. There may be no accurate list of the animals and plants that have been exterminated by the introduction of rats and mice, goats, rabbits, mongooses, and other alien organisms. House cats are examples of particularly lethal animals that accompany humans and do irreparable harm to local faunas.

Outright, specific, destruction of other organisms is usually achieved by some combination of technology and relatively small target populations. Technology can affect the risk to the hunter. When the quarry is a black rhino, the risk to the hunter is probably greater when the weapon is a bow and arrow or spear than when it is a high-powered rifle.

A Critical Localized Resource

Organisms for whom a special, limited, and localized resource is critically important are extremely vulnerable to human and other activities. For example, tropical beaches that attract tourists from the developed world are often important nesting sites for sea turtles. Relatively small stretches of beach may be crucial to the survival of turtle populations from large areas. Other examples are provided by hibernation sites used by animals from large areas. Appropriate examples would include the sites used by monarch butterflies from North America (*Danaus plexippus*) or the caves and mines that serve as overwintering sites for many species of temperate bats.

In these situations, adequate basic knowledge about a species is essential to ensure that human activities do not destroy the localized resource and thus the organisms that use it. Important sites, especially those of a relatively small scale, may easily be overlooked. For example, Fenton et al.[21] observed large numbers of bats visiting a small (1 m diameter) pool in rainforest in Mount Revelstoke National Park. Why the bats visited the pool remains unknown, but what is more disconcerting is the fact that the importance of this site could easily have remained unnoticed.

Bizarre Interactions

The network of global ecosystems provides astonishing evidence of wide-ranging interconnections. There are many animals and plants for which we have nothing more than a description and a name. Increasing knowledge provides astonishing examples of intricate relationships between species. How easy it could be to destroy an interaction by removing part of it.

For example, in Africa two species of ox-peckers (*Buphagus* spp.), birds related to starlings, make their living by gleaning ticks and other ectoparasites from large mammals. Today, ox-peckers are rarely seen outside large game reserves such as national parks, even though cattle should provide excellent feeding stations. Cattle ranchers in Africa typically dip their animals to control tick infestations and associated problems. A diet of insecticide-laden ticks is lethal to ox-peckers. The connection is obvious and explains the decline of ox-pecker populations throughout Africa. In Zimbabwe, successful reintroductions of ox-peckers have coincided with farmers avoiding toxic dips.[22]

In parts of Africa some birds known as honeyguides (*Indicator* spp.) lead humans and honey badgers (*Mellivora capensis*) to bee hives. In Kenya, Boron people who gather honey are much more efficient at finding hives when they use the birds than when they work alone.[23] Increasing urbanization may mean decreasing honey badger populations and less human dependence upon wild honey. This could extinguish the relationship between the honeyguides and other animals, a relationship that is an astonishing example of interspecific communication.

Fragmentation of Populations

Biologists presume that gene flow, the transfer of genetic information between generations, is one of the glues that holds together the integrity of species. The real importance of gene flow in this role has been questioned,[24] but some studies have shown how quickly genes can move through populations.[25] If gene flow is important in maintaining species integrity, barriers to gene flow will lead over time to reproductive isolation and, eventually, to new species.

Many human activities affect habitat structure. For example, clearing forests for agricultural purposes can produce a combination of forests and fields where forests once predominated. Isolated patches of forest are often interconnected by corridors such as water courses or fence rows. In eastern Ontario, this landscape mosaic has important consequences for animals and plants. Middleton and Merriam[26] concluded that woodland plants and animals have effective short-range dispersal mechanisms that maintain heterogeneity in the face of severe disruption. Freemark and Merriam[27] found that a combination of woodlot area and its tree species composition affected the species of birds living there.[27]

One current view is that many populations, particularly in heterogeneous habitats, are spatially and functionally discrete units known as metapopulations.[28] At what point does a metapopulation become a race? A variety? A subspecies? What features act as barriers between populations? For animals that cannot fly, mountains and rivers are traditional barriers to movement.[29] In habitats modified by humans, roadways significantly interfere with the movements of small forest mammals (Figure 5.3) but not with grassland species.[30] Cultivated fields also interrupt the movements of small forest mammals.[31]

One point of view is that gene flow contributes to genetic variation and the health of populations and that healthy populations should be varied ones. An event that reduces populations to very low levels can have a bottleneck effect: in small, inbreeding populations, deleterious, genetically controlled factors may become fixed, rapidly reducing the chances of survival. The Florida panther may be an example of this kind of situation. The lack of genetic variability in elephant seals (*Mirounga* spp.)[32] and cheetahs (*Acinonyx jubatus*)[33] is often attributed to bottlenecks. In cheetahs, low genetic variation does not coincide with reduced morphological variation.[34] The Indian rhinoceros (*Rhinoceros unicornis*) provides an alternative example. Although in 1962 the population of Indian rhinos in the Chitwan Valley of Nepal was reduced to 60–80 animals, the current population of about four hundred animals shows a high level of genetic variability,[35] perhaps reflecting the legacy of a huge original population.

In some cases a high degree of genetic variation is associated with very large populations. Mexican free-tailed bats provide one example, and here there is no evidence of genetic isolation of the more northern populations.[36] The diversity of living organisms makes it easy for one to select examples supporting a particular point of view with regard to the relationship between genetic variation and the success of populations.

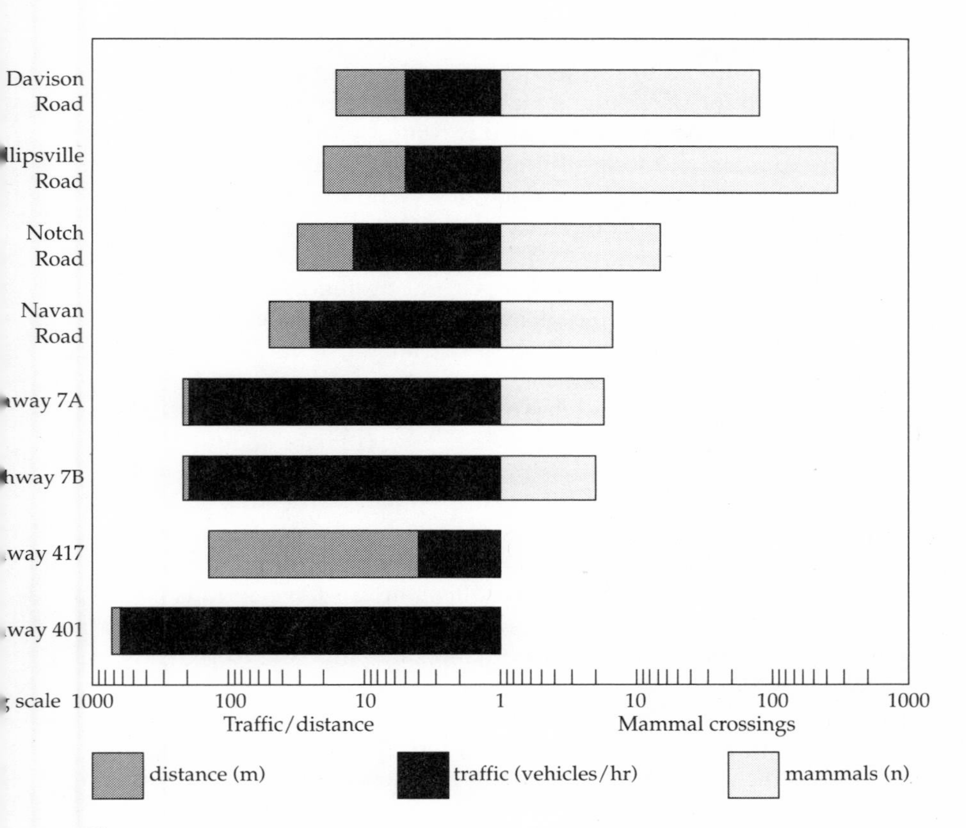

Figure 5.3
Roads can be barriers to the movements of small forest mammals, as illustrated by the study by Oxley et al. (1974) of the numbers of small forest mammals crossing roads of different widths and traffic volumes in Ontario. The study findings indicate that road clearance rather than traffic volume determines the effectiveness of a roadway as a barrier to the dispersal of small forest mammals. Davison Road and Phillipsville Road are small, single-lane dirt roads, while Notch Road and Navan Road are two-lane paved secondary roads. Highways 7A and 7B are two-lane highways, and Highways 401 and 417 are divided, four-lane highways where sampled. Distances (in metres) reflect the width of the cleared roadway and shoulder (30 m for Highway 7, 100 m for 401 and 417). Traffic volume is represented in vehicles per hour; since Highway 417 had been built but was not yet open at the time of the study, it was little used in the study.

Habitat Destruction

Habitat destruction probably accounts for more species impoverishment than any other factor. Furthermore, habitat destruction often exterminates species as yet unknown to science. As I have noted, unnamed species provide no rallying point for conservation efforts.

Quite simply, few organisms can survive when their habitats are destroyed. Many natural phenomena destroy large areas of habitat. The 1986 eruption of Mount St Helens is one example. Intense tropical storms, such as the cyclones that devastated parts of Bangladesh in 1991 are another. Although humans are part of the natural system, their capacity for large-scale habitat destruction is almost unrivalled. The accelerating growth of the global human population is both a direct and an indirect cause of habitat destruction. Human beings have used technology to harvest natural systems. Clear-cutting of forests is a glaring example of intense habitat destruction. Whether the clearcut occurs in eastern or western Canada, in the Amazon basin, in the Philippines, or in Uganda, the net effect is the same on the organisms of the forest: total destruction. The flooding associated with hydroelectric projects also achieves high levels of habitat destruction.[37]

But humans destroy habitats in other ways too. Urban sprawl and other pressures of increasing human populations destroy vast areas of habitat. Forests and woodlands are often under pressure from the demand for land for cultivation, materials for building and communication, and firewood. Firewood collection can have serious consequences for woodlands, even when fallen timber is the only material removed. In open, deciduous woodlands, like mopane woodlands in southern Africa, fallen timber prevents leaves that have been shed from blowing away. This process may be vital for keeping biomass in the system and for preventing erosion. But heat for cooking and warmth directly affect human health and standards of living, bringing us back to questions of population, energy consumption, and lifestyle, as is discussed elsewhere in this volume.

Since, as we have noted, even small-scale habitat destruction can directly affect the integrity of populations, it is evident that stopping large-scale habitat destruction is one of the first measures that must be taken to alleviate the problem of species impoverishment.

CONSERVATION LANGUAGE

In the early 1960s the International Union for the Conservation of Nature and Natural Resources (IUCN) took action to identify organisms whose future was in jeopardy. In 1966 the IUCN published "The Red Data Book," which addressed the status of species of mammals on a worldwide basis. Subsequently, many other "red books" have been published, addressing the status of species on local, national, and global scales. The red books often have had a strong positive effect, particularly when the information is used in concert with initiatives from CITES, the Convention for International Trade in Endangered Species.

For example, in South Africa eight red book reports had been published by 1986, covering birds, small mammals, fishes, large mammals, reptiles and amphibians, and vascular plants. Since the 1976–77 publication of the South African Red Data Books no known species have become extinct in that political area. Furthermore, populations of Cape mountain zebra (*Equus zebra zebra*) and of white rhino have recovered substantially.[38]

In its red data books, the IUCN categorizes species by their status as follows:

- *Endangered species* are those in immediate danger of extinction.
- *Vulnerable species* are those about to move into the endangered category because of continued reduction in numbers.
- *Rare species* are not in immediate danger, but low numbers or extreme habitat specialization make them very vulnerable.

IUCN initiatives, coupled with the actions of CITES, may have been instrumental in preserving many species that otherwise would have disappeared. But there are four problems with this approach to conservation: (1) unnamed and unknown species; (2) lack of data; (3) inaccurate or disputed data; and (4) public perception.

Unnamed, Unknown Species

As noted earlier organisms not yet described are not protected by these measures. Furthermore, misleading taxonomy can make some organisms vulnerable. For example, although three species of tuatara (*Sphenodon*), large iguana-like reptiles, were originally described from a few islets off the coast of New Zealand, for some reason only one is named in the 1895 Animals Protection Act in New Zealand.[39] Recent work has confirmed that although one species of tuatara is extinct, two survive.[40] One of the surviving species remains unrecognized in law and its numbers have dwindled to fewer than three hundred animals.

Lack of Data

A lack of information about most organisms – their numbers, distribution, and habits – makes it impossible to classify them accurately according to the IUCN terminology. For example, about nine hundred species of bats are known to science,[41] but for most there is not enough information about population size or biology to permit intelligent application of the IUCN labels. In Canada, where twenty-two species occur regularly,[42] only spotted bats and pallid bats (*Antrozous pallidus*) may qualify as "rare." The Canadian range of these species is limited to a small area in the southern Okanagan Valley of British Columbia. In 1979, two pallid bats were caught in the southern Okanagan,[43] and, in spite of continued searching, no more were encountered until the summer of 1990.[44] Should pallid bats, apparently rare in Canada, be protected when they are often very common in the main parts of their range?

Of nine hundred or so species of bats in the world, there may be ten South Pacific island species of fruit bat or flying fox (*Pteropus* spp.) that are endangered. In these cases, small island populations have been hunted for years by local people and consumed as food on festive occasions. When the implements of the hunt changed from nooses on the ends of long poles to shotguns, the harvesting pressure on the bats increased and their populations plummeted.[45] In this instance, local harvesting, which is beyond the control of the IUCN and CITES, has had a devastating impact. CITES, however, by banning trade in flying foxes, can reduce the impact of harvesting in other jurisdictions. Most species of plants and animals are like most species of bats: our ignorance makes the IUCN labels meaningless.

Inaccurate or Disputed Data

Inaccurate or disputed data about the population and status of species also can pose difficult problems for conservationists. One fundamental difficulty is the accuracy with which humans measure the size of animal populations. A second reality is that the numbers of organisms may differ substantially from one place to another. The status of African elephants (*Loxodonta africana*) illustrates these points (Figure 5.4).

The recent controversy over the decision to include African elephants (*Loxodonta africana*) in CITES' Appendix 1 was made at the urgent request of countries in east Africa, over the very strong objections of countries in southern Africa. A listing in Appendix 1 bans all legal trade in the products of the listed animal (in this case ivory and skin). In east Africa, extensive poaching has reduced African elephant populations to very low levels, while in southern Africa, culling operations are necessary to keep large elephant populations at levels supportable by the habitats in which they occur. In some southern African countries, proceeds from the sale of elephant products directly finance conservation programs.[46]

"How many African elephants are there in the wild?" is a central question in this issue. Cumming et al.[47] present evidence that African

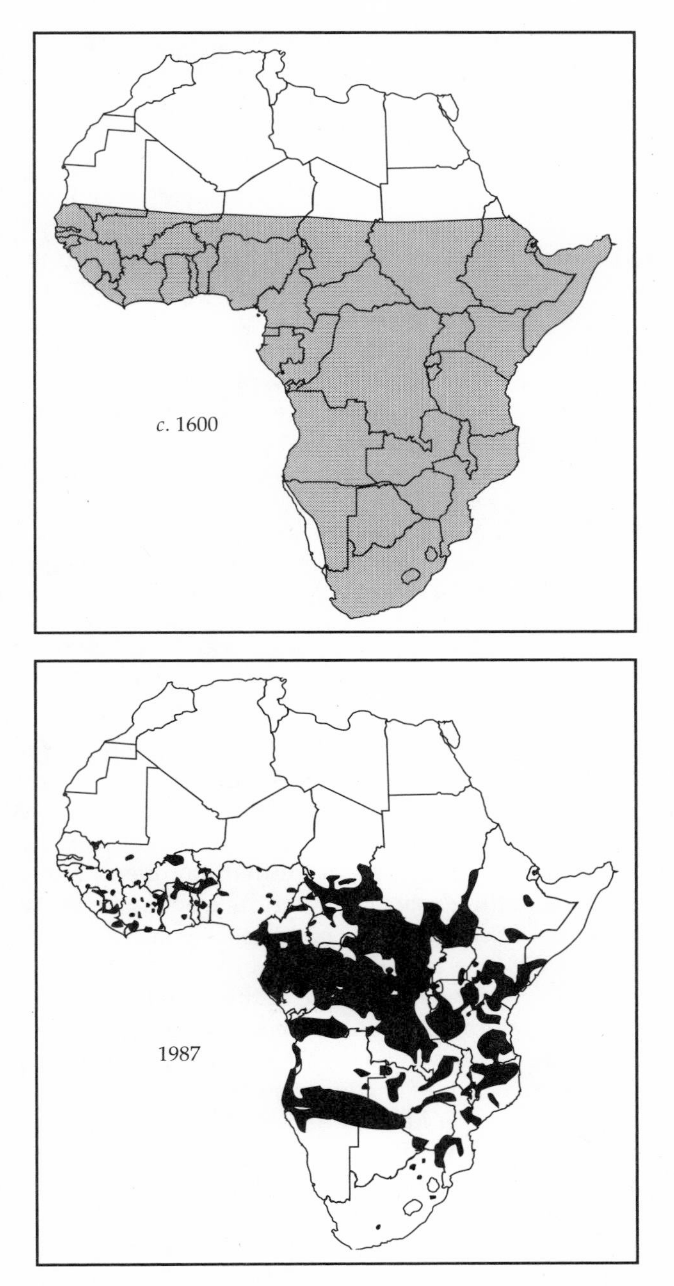

Figure 5.4
The geographic range of African elephants *c.* 1600 and in 1987. (After Cumming et al. 1990.)

elephants are not endangered over their range, although in some areas their numbers have declined dramatically. The alternative point of view, vigorously expressed by some prominent conservationists, is that African elephants are close to extinction. A widespread publicity campaign conducted in Canada and the United States during 1990 presented the east African point of view and gained widespread popular support. Ivory poachers kill many elephants every year and the hypothesis presently accepted by CITES is that listing African elephants in Appendix 1 should drastically reduce trade in ivory by making it illegal.

The situation, which is discussed broadly by Chadwick,[48] is far from being resolved. The controversy typifies some of the problems inherent in making decisions when the data base is weak. CITES and IUCN measures are meaningless if there is no local protection of the animals in question. There has been similar controversy about the status of leopards (*Panthera pardus*). Leopards had been included on CITES Appendix 1, but some countries were allowed exceptions in the form of trade quotas. These exceptions were granted for countries showing effective management of leopards; Botswana was the first to be granted a trade quota.[49]

Public Perception

The IUCN labels have given the public (broadly defined) a means of making certain organisms the focus of concern and action. When speaking to public audiences or naturalists' groups about bats I am frequently asked to identify species that are endangered. If I reply that we do not know enough to make that judgment, many people seem to relax, assuming that if they are not endangered, then they do not warrant further concern. The long term success of conservation initiatives may be adversely affected by such attitudes.

There are other cases where the use of the wrong label can be fatal. The case of the tuataras has already been noted. But imagine suggesting in 1800 to settlers in eastern North America that passenger pigeons would be extinct in just over one hundred years. This species was so abundant that the idea would have been thought preposterous. Passenger pigeons had very slow rates of reproduction (one egg per year), giving them a very low r value. The rates at which passenger pigeons were harvested exceeded this species' reproductive capacity to replace harvested individuals. In other words, the rates of harvesting exceeded r and extinction was just a matter of time.

In much of North America, big brown bats (*Eptesicus fuscus*) and little brown bats (*Myotis lucifugus*) commonly roost in buildings.

These two species illustrate two important points about public attitude and conservation. The first is image. Bats are not welcome in most houses, for they are thought to be dangerous because of rabies or histoplasmosis. Bats are susceptible to rabies, but they do not "carry" it and there are no records of histoplasmosis from bats in Canada.[50] House bats can be a nuisance, but they are not dangerous to people. In cases such as these the public perception of "good" and "bad" organisms affects their chances of survival.

The second difficulty is the general public's view that since these bats appear to be common *and* they live in buildings there is no need to conserve them. People often can evict bats from buildings by excluding them, although in the United Kingdom this is unlawful.[51] What happens to displaced bats? While big brown bats tend to move next door,[52] little brown bats seem to disappear from the population. For them, eviction may be a death sentence.[53] In the United Kingdom and some other countries in Europe, house bats are protected by law.[54]

In short, the public's perceptions about which organisms deserve concern and action can have as many negative as positive effects. One challenge to scientists and conservationists is to convince the public that for most organisms we need more information before we can identify effective conservation plans. This problem is exacerbated by the fact that many conservation organizations fund only those projects involving organisms already labelled as rare or endangered.

Another challenge is to make the public appreciate that even animals and plants they do not *like* deserve attention. It is important to realize that as long as our knowledge about species and ecosystems is incomplete, mistakes in conservation plans will drive some species to extinction.

SELFISH SOLUTIONS

Putting a stop to species impoverishment is everybody's problem, but many people need to be convinced of this. Perhaps it is easy to be complacent about the problem when we are bombarded with stories about all of the conservation programs that are under way. For example, many zoos are running captive breeding programs designed to protect species threatened with extinction. However, such programs are often fraught with difficulty, such as the introduction into the wild of diseases contracted by animals in captivity. In one instance this was narrowly averted when lion tamarins (*Leontopithecus rosalia*) about to be returned to their native habitat in Brazil were found to be harbouring an "emerging virus."[55] Clearly, the presence of disease can profoundly affect efforts to conserve species.[56]

Like all other animals whose behaviour has been studied in any detail, human beings are selfish. It is easy to find examples of animal behaviour motivated by self-interest, and even behaviours that appear to be altruistic are best explained as self-interested acts.[57] Sceptics about the fundamental pervasiveness of selfishness should consider Wilkinson's data on the common vampire bat's (*Desmodus rotundus*) sharing of food, in this case blood.[58] Here, an apparently selfless act carries at least as many long-term benefits for the donor as for the recipient. Regurgitating some blood for a roost-mate that did not feed costs the donor relatively little but means survival to the recipient. Since on any night any adult common vampire bat has a 7 per cent chance of not feeding, today's donor will be a recipient at least once a month. In common vampire bats, sharing food is a selfish business. How can we make habitat preservation the selfish business of people?

Habitat destruction is the main cause of species impoverishment, indiscriminately causing the demise of endangered, common, little-known, and unknown organisms. Stopping habitat destruction should be the first priority. Making the intact habitat worth more than the land without the habitat can be a practical way to achieve this goal. There are two promising developments in this area: ecotourism, and game management.

The rainforest, the desert, the wilderness, and so forth have become very trendy vacation destinations today. When the area to be visited is in the Third World, tourist activity can produce local employment and bring in much needed foreign currency. When the area is in a remote part of Canada, for example, the tourist traffic can have a similarly stimulating effect on the local economy. In some cases the goal of tourists is to see particular animals, such as gorillas, elephants, rhinos, or whales. In other cases, tourists go to see the habitat – the rain forest, desert or wilderness – itself.

Inherent in the ecotourist approach is the conservation of tracts of natural habitat. Whether the focus of the tourism is the habitat or some of the organisms that live there, the net effect is the same. By preserving the habitat the organisms that live there are protected.

Wildlife management is another strategy that leads to the preservation of large areas of natural habitat. Inherent in wildlife management, however, is the harvesting, i.e. killing, of wildlife, sometimes for protein, but more often for trophies. This approach is viable only where there is game to harvest, and some animals have more value as trophies than others. One may want to display the head of a lion while most of a rabbit, even its head, may end up in the pot.

Since wildlife management usually entails killing animals, its value as a conservation tool is often underestimated by the average ecotourist. Cumming[59] has provided data about the economic benefits of game ranching in Zimbabwe, illustrating that the gross return in Zimbabwe dollars is 18 cents per kilogram of game harvested as opposed to 6 cents per kilogram of cattle harvested. When the main focus of the harvesting operation is trophies, local people benefit from employment and from the byproducts of the hunting operation – mainly meat. Again, the important point is that the habitat in which the quarry occurs becomes more valuable intact, giving local people a strong motivation to preserve it. Furthermore, in a country like Zimbabwe big game hunting brings in much-needed foreign currency.

Ecotourism and game management appeal to the selfish motives of the local inhabitants as well as to those of people living elsewhere. Game management does not need to conflict with the goals of ecotourists, as many safari outings collect photographic trophies rather than parts of dead animals. Both of these approaches, however, bring problems such as (1) assessing the impact of tourist activity on the target habitat/organisms, (2) determining the sizes of areas that must be set aside, (3) dealing with perceptions of groups concerned about the exploitation of animals.

Effects of Tourism

What effect do tourists and tourist operations have on habitats and organisms? The scale of such operations has an immediate impact, whether this is related to sewage or garbage disposal, or to supplying the transient and resident populations of humans with food and other services. Biologists should be challenged to identify organisms that are indicators of disturbance. The responses of these indicators can be used to set limits to disturbance that ensure that the carrying capacity of the system is not exceeded. In general, smaller organisms may be better indicators of disturbance than larger ones. A survey in the Yucatan in January 1991 revealed that some kinds of bats were good indicators of levels of habitat destruction in the form of deforestation.[60] Where specific species are the target of the conservation effort, we must know what levels of disturbance they will tolerate (e.g., mountain gorillas, *Gorilla gorilla*).[61]

The Size of Protected Areas

In mammals there is a general correlation between the size of the body and the extent of the home range;[62] for example, elephants use much larger areas than mice. Animals that are territorial, defending their home ranges against conspecific intruders, usually occur at lower population densities than those which live in large groups; however, species may change their patterns of land use from season to season[63] or from place to place.[64]

It is clear that tracts of protected habitat must be large enough to support breeding populations of a good size. Small Ontario provincial parks are rarely home to large mammals such as white-tailed deer (*Odocoileus virginianus*), black bears (*Ursus americanus*) or bobcats (*Felis rufus*). The same pattern persists elsewhere; Freemark and Merriam, for example, have demonstrated that forest patch size and composition affects the species composition of bird communities in woodlots.[65] Population models may be used to estimate preferred sizes of habitat tracts. For example, Lande[66] has shown that many of the present or proposed forest reserves in western North America are not large enough to ensure the survival of spotted owls (*Strix occidentalis caurina*).

In Zimbabwe, land areas of at least 130 km^2 are necessary to serve as the base for big game safaris in which the quarries include African elephants, African lions and Cape buffalo (*Syncerus caffer*). On smaller holdings, other game can be hunted. Areas of at least 60 km^2 can serve for plains game (mainly antelope) provided that no domestic stock are present. Game ranches, per se, are holdings of less than 60 km^2 where domestic stock and game share the available habitat.[67]

In cases where some feature of the environment is a critical resource for organisms, the preserved area must encompass the vital feature.

Public Opinion

Today, many people object strenuously and even violently to any exploitation of animals. Animal rights activists can influence public opinion and affect the outcome of conservation operations far beyond where they live and work. Evidence of this is provided by the impact in Newfoundland and some other parts of Atlantic Canada of the European reaction to the harvest of harp seals (*Phoca groenlandica*). These situations are often coloured by the choice of words: "killing" versus "harvesting"; "seal pups" versus "baby seals."[68]

As many (but not all) southern African wildlife biologists and conservationists have noted, placing African elephants on CITES' Appendix 1 may provide another example of remote interference. In southern African countries, the proceeds of the sale of elephant products are critical to national conservation efforts. A relevant question, then, is whether it is possible to distinguish legal ivory from poached ivory. The ban on elephant products may directly affect the safari hunting operations in southern African countries.

Placing leopards on CITES Appendix 1 can have a similarly negative impact on the game management scene in African countries. Martin and de Meulenaer[69] reported that there was considerable evidence that sustained, low-level harvesting of leopards did not jeopardize the populations. Indeed, they noted that efforts to eradicate leopards from areas with cattle have not succeeded and that leopards occur even in some large African cities. They pointed out that if human population trends in Africa hold their present course over the next twenty years, the amount of suitable habitat available to leopards will decrease by 50 per cent. Moreover, harvesting is not the insidious and immediate threat to the survival of leopards that is presented by increased habitat destruction as a result of a growing human population.

SUMMARY

These considerations have brought us back to questions of habitat destruction, the human population, and our attitude to the global system. Remote solutions to conservation problems do not work. Idealistic academic arguments, however articulate,[70] will not ensure the protection of appropriate tracts of habitat, but a local interest in preservation can.[71] This means that efforts to save flying foxes in Guam will be effective only if they are supported by local citizens. This, in turn, probably means harvesting them at appropriate levels. Will we see hunting in national parks as salvation or as sacrilege?[72] It may also mean that saving black rhinos will involve some licensed hunting, with the licences being extremely expensive. It is possible (or likely) that in the next five years South Africa and Zimbabwe will each sell one licence for black rhino each year.[73] The proceeds from the hunt will be used to support conservation and the high price for the licences will make the rhinos worth more alive than dead.

Martin and Taylor[74] have presented a model pattern of land use for the Sebungwe area of Zimbabwe. In this model, national parks serve as refuges and as food sources for wildlife, while adjoining safari hunting areas support subsistence farming and wildlife exploitation. Beyond the safari areas are lands supporting higher levels of agricultural and other activities (Figures 5.5 and 5.6). The patterns of land use are generally in tune with the carrying capacity of the land. The local people have an immediate stake in, and benefit from, the tracts of land where natural habitats are preserved. The habitats themselves bring foreign visitors, jobs, and money and may positively affect local standards of living.

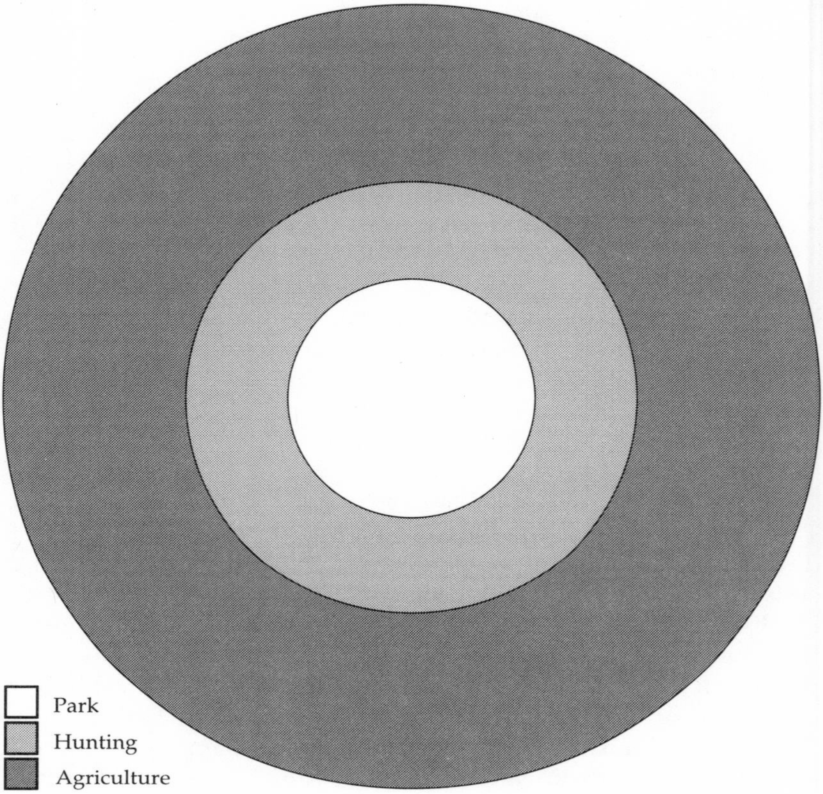

Park
Hunting
Agriculture

Figure 5.5
A model for land use. Parks and reserves serve as resource areas for wildlife; these are bounded by areas for hunting and subsistence farming, which are in turn surrounded by areas of more intensive land use.

Progressing species impoverishment is one reflection of the continued deterioration of the global ecosystem. Stopping large-scale habitat destruction can slow this process and it is encouraging that there is ample evidence of the recuperative powers of both species and ecosystems. Providing solutions that appeal to the selfishness of people seems to offer the best chance for a long-term halt to species impoverishment.

Is it too late? Already some biologists are starting to identify areas that need immediate protection. The Rapid Assessment Program (RAP) is one way to obtain a quick evaluation.[75] Meanwhile, May[76] points out that others have begun to think of making value judg-

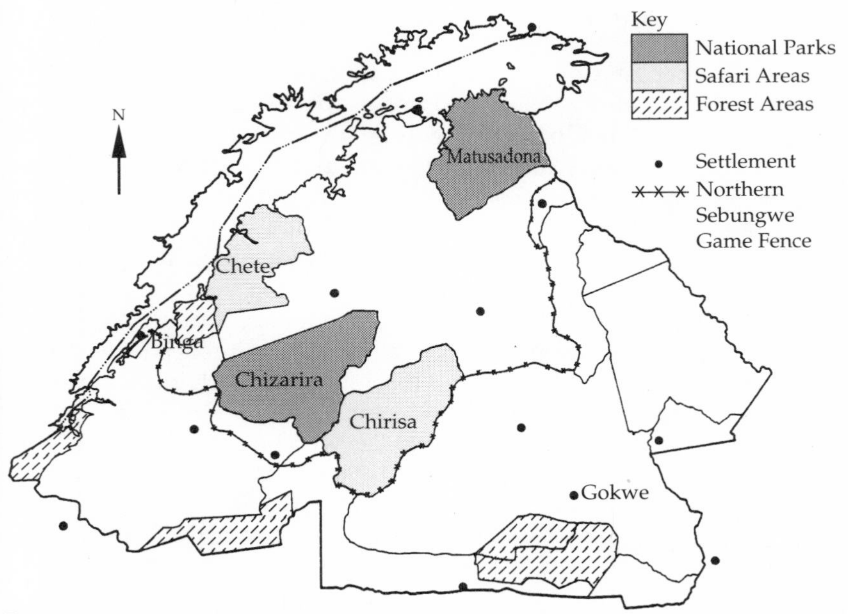

Figure 5.6
Existing parks and safari areas in the Sebungwe district of Zimbabwe are proposed as the basis for implementing the model for land use shown in Figure 5.5. (After Martin and Taylor 1983.)

ments about which organisms are most worthy of protection. Should the United States and Canada continue to invest large amounts of resources in the protection of a small population of whooping cranes (*Grus americana*)? Would this money be better spent setting aside and protecting large tracts of habitat?

At the root of the problem is the attitude of humans to themselves, to other humans, and to the rest of the world. Some people have suggested that the Christian view of the world – that it was made by God and put at our disposal – is at least partly responsible for massive habitat destruction.[77] I think that habitat destruction and disruption is a much more pervasive problem, and that there is evidence to support this view.[78] It reflects the attitude that being able to afford to do something is the necessary justification for doing it. Some environmental education programs invoke a religious view that may be more acceptable in some societies than in others.[79] As Parker has repeatedly noted,[80] humans must now reassess their relationships with and obligations to the environment.

ACTION

If even 10 per cent of the people in Canada adopted lifestyles that consumed less energy we could see a positive effect on the global situation. Our selfish benefit would be lower energy bills. Recognizing that we are part of the problem immediately puts part of the onus on us to effect change. To succeed in this venture, we must stop using guilt and blame when speaking of environmental problems. Our selfishness is a more powerful force in favour of making positive changes than guilt about what we have done in the past. Allocating blame, to the media, to politicians, to business, or to our neighbour, exonerates us from responsibility. A positive approach is to accept responsibility for our actions and recognize that each of us can make a contribution.

The key to our continued survival lies with our young people, who need positive things to do to counteract a climate of despair. Actions, however, must include careful shunning of true believers and fanatics, as well as of finger-pointing accusations about the behaviour of other people. In the words of Pogo, "We have met the enemy and he is us." To this I can add: "We have seen the solution and it is us."

ACKNOWLEDGMENTS

I thank the Royal Society of Canada and Queen's University for the opportunity to participate in the Planet Earth symposium and Professors J.H. Spencer and R.A. Price for organizing the event. Other symposium speakers and members of the audience have broadened my view of ecological problems and solutions. Several colleagues took the time to review earlier drafts of this chapter, offering critical and constructive suggestions and sharing their views. For their efforts in this context I am very indebted to Lalita Acharya, Leesa Fawcett, G. Brian Golding, M. Brian C. Hickey, Rowan B. Martin, H. Gray Merriam, Marilyn E. Scott, Geoffroy G.E. Scudder, and Clive M. Swanepoel. My research on bats has been supported by the Natural Sciences and Engineering Research Council of Canada. My wife Eleanor continues to tolerate my sojourns to collect data about bats in different parts of the world.

NOTES

1 Risebrough, R.W., Walker, W., Schmidt, T.T., Delappe, B.W., Conners, C.W. 1976. Transfer of chlorinated biphenyls to Antarctica. *Nature* 264: 738–9.

2 Mayr, E. 1940. *Animal Species and Evolution.* Cambridge, Mass: Harvard University Press.

3 Raup, D.M. 1986. Biological extinction in earth history. *Science* 231: 1528–35.

4 Mayr 1940.

5 Van Zyll de Jong, C.G. 1985. *Handbook of Canadian Mammals.* 2. Bats. Ottawa: National Museums of Canada.

6 Wilkins, K.T. 1989. Tadarida brasiliensis. *Mammalian Species* 331: 1–10.

7 O'Brian, S.J., Mayr, E. 1991. Bureaucratic mischief: recognizing endangered species and subspecies. *Science* 251: 1187–8.

8 Fergus, C. 1991. The Florida panther verges on extinction. *Science* 251: 1178–80.

9 O'Brian and Mayr 1991.

10 Ibid.

11 Fergus 1991; Gittleman, J.L., Pimm, S.L. 1991. Crying wolf in North America. *Nature* 351: 524–5; Wayne, R.K., Jenks, S.M. 1991. Mitochondrial DNA analysis implying extensive hybridization of the endangered red wolf, *Canis rufus. Nature* 351: 565–8.

12 Considine, T.J., Frieswyk, T.S. 1982. Forest statistics for New York 1980. U.S. Department of Agriculture, Forest Service, Northeastern Station, New York State Department of Environmental Conservation, Resource Bulletin NE-71; Considine, T.J. 1984. An analysis of New York's timber resources. U.S. Department of Agriculture, Forest Service, Northeastern Forest Experimental Station, Resource Bulletin NE-80.

13 Hammond, N. 1982. *Ancient Mayan Civilization.* New Brunswick, N.J.: Rutgers University Press.

14 Garlake, P. 1985. *Great Zimbabwe Described and Explained.* Harare: Zimbabwe Press.

15 Krebs, C.J. 1972. *Ecology, the Experimental Analysis of Distribution and Abundance.* Toronto: W.B. Saunders.

16 Sudman, S., Sirken, M.G., Cowan, C.D. 1988. Sampling rare and elusive populations. *Science* 240: 991–6.

17 Trivers, R. 1985. *Social Evolution.* Philadelphia, Pa.: Benjamin/Cummings.

18 Colinvaux, P. 1978. *Why Big Fierce Animals are Rare: An Ecologist's Perspective.* Princeton University Press.

19 Herero, S. 1985. *Bear Attacks, Their Causes and Avoidance.* New York: Nick Lyons Books.

20 Cumming, D.H.M., Du Toit, R.F., Stuart, S.N. 1990. African elephants and rhinos: status survey and conservation action plan. IUCN/SSC African Elephant and Rhino Specialist Group. IUCN Gland, Switzerland.

21 Fenton, M.B., Merriam, H.G., Holroyd, G.L. 1983. Bats of Kootenay, Glacier, and Mount Revelstoke National Parks in Canada: identification

by echolocation calls, distribution, and biology. *Canadian Journal of Zoology* 61: 2503–8.

22 R.B. Martin, National Parks and Wildlife, Zimbabwe. Letter to author, May 1991.

23 Isack, H.A., Reyer, H.-U. 1989. Honeyguides and honey gatherers: interspecific communication in a symbiotic relationship. *Science* 195: 82–4.

24 Erlich, P.H., Raven, P.H. 1969. Differentiation of populations. *Science* 165: 1228–32.

25 Baker, A.E.M. 1981. Gene flow in house mice: introduction of a new allele into free-living populations. *Evolution* 35: 243–58.

26 Middleton, J., Merriam, H.G. 1983. Distribution of woodland species in farmland woods. *Journal of Applied Ecology* 20: 625–44.

27 Freemark, K.E., Merriam, H.G. 1986. Importance of area and habitat heterogeneity to bird assemblages in temperate forest fragments. *Biological Conservation* 36: 115–41.

28 Levins, R. 1970. Extinction. In *Some Mathematical Questions in Biology: Lectures on Mathematics in the Life Sciences*, ed. M. Gerstenhaber. Providence, R.I.: American Mathematical Society.

29 Mayr 1940.

30 Oxley, D.J., Fenton, M.B., Carmody, G.R. 1974. The effects of roads on populations of small mammals. *Journal of Applied Ecology* 11: 51–9.

31 Merriam, H.G., Kozakiewicz, M., Tsuchiya, E., Hawley, K. 1989. Barriers as boundaries for metapopulations and demes of *Peromyscus leucopus* in farm landscapes. *Landscape Ecology* 2: 227–35.

32 Bonnell, M.L., Selander, R.K. 1974. Elephant seals: genetic variation and near extinction. *Science* 184: 908–9.

33 O'Brian, S.J., Roelke, M.E., Marker, J., et al. 1985. Genetic basis for species vulnerability in the cheetah. *Science* 227: 1428–34.

34 Kierser, J.A., Groeneveld, H.T. 1991. Fluctuating odontometric asymmetry, morphological variability and genetic monomorphism in the cheetah, *Acinonyx jubatus*. *Evolution* 45: 1175–83.

35 Dinerstein, E., McCracken, G.F. 1990. Endangered greater one-horned rhinoceros carry high levels of genetic variation. *Conservation Biology* 4: 417–22.

36 McCracken, G.F. 1988. Genetic variation in the bat *Tadarida brasiliensis mexicana* and in *Rhinoceras unicornis*, a large, non-volant mammal. *Bat Research News* 29: 49.

37 Richardson, B. 1975. *Strangers Devour the Land: The Cree Hunters of James Bay versus Premier Bourassa and the James Bay Development Corporation*. Toronto: Macmillan.

38 Smithers, R.H.N. 1986. *South African Red Data Book: Terrestrial Mammals*. South African National Scientific Programmes Report no. 125.

39 May, R.M. 1990. Taxonomy as destiny. *Nature* 347: 129–30.

40 Daugherty, C.H., Cree, A., Hay, J.M., Thompson, M.B. 1990. Neglected taxonomy and continued extinctions of tuatara (*Sphenodon*). *Nature* 347: 177–9.
41 Fenton, M.B. 1983. *Just Bats*. University of Toronto Press.
42 Van Zyll de Jong 1985.
43 Fenton, M.B., Van Zyll de Jong, C.G., Bell, G.P., Campbell, D.B., Laplante, M. 1980. Distribution, parturition dates and feeding of bats in south-central British Columbia. *Canadian Field-Naturalist*, 94: 416–20.
44 R.M. Brigham, Department of Biology, University of Regina. Letter to author, May 1991.
45 Flannery, T.F. 1989. Flying foxes in Melanesia: populations at risk. *Bats* 7(4): 5–7; Wiles, G.J. 1990. Giving flying foxes a second chance. *Bats* 8(3): 3–4.
46 Martin, R.B., Craig, G.C., Booth, V.R. (eds) 1989. *Elephant Management in Zimbabwe*. Department of National Parks and Wild Life Management, Harare. Pretoria: Haum.
47 Cumming et al. 1990.
48 Chadwick, D.H. 1991. Elephants: out of time, out of space. *National Geographic* 179(5): 2–48.
49 Martin, R.B., de Meulenaer, T. 1988. Survey of the status of leopard (*Panthera pardus*) in sub-Saharan Africa. Ottawa: CITES.
50 Brigham, R.M., Fenton, M.B. 1987. The effect of roost sealing as a method to control maternity colonies of big brown bats. *Canadian Journal of Public Health* 78: 47–50.
51 Fenton 1983.
52 Brigham and Fenton 1987.
53 Neilson, A.L. 1991. Population ecology of the little brown bat, *Myotis lucifugus*, at the Chautauqua Institution, Chautauqua, New York. M.Sc. Thesis, Department of Biology, York University, North York, Ontario.
54 Corbet, G.B., Harris, S. (eds). 1991. *The Handbook of British Mammals*, 3rd ed. Oxford: Blackwell Scientific Publications.
55 Anderson, C. 1991. Emerging virus threat. *Nature* 351: 89.
56 See, for example, Scott, M.E. 1988. The impact of infection and disease on animal populations: implications for conservation biology. *Conservation Biology* 20: 40–56.
57 Trivers 1985.
58 Wilkinson, G.S. 1985. The social organization of the common vampire bat. I. Pattern and cause of association. *Behavioral Ecology and Sociobiology* 17: 111–21.
59 Cumming, D.H.M. 1989. Commercial and safari hunting in Zimbabwe. In R.J. Hudson, K.R. Drew and L.M. Basking, eds, *Wildlife Production Systems: Economic Utilisation of Wild Ungulates*. Cambridge University Press.
60 Fenton, M.B., Acharya, L., Audet, et al. 1992. Phyllostomid bats (Chiroptera: Phyllostomidae) as indicators of habitat disruption. *Biotropica*.

61 Mowat, F. 1987. *Virunga, the Passion of Dian Fossey.* Toronto: McClelland and Stewart.
62 Vaughan, T.A. 1986. *Mammalogy,* 3rd edn. Philadelphia, Pa.: Saunders College Publishing.
63 Wilson, E.O. 1975. *Sociobiology, the New Synthesis.* Cambridge, Mass: Harvard University Press.
64 See, for example, Brigham, R.M. 1991. Flexibility in foraging and roosting behaviour by the big brown bat (*Eptesicus fuscus*). *Canadian Journal of Zoology* 69: 117–21.
65 Freemark and Merriam 1986.
66 Lande, R. 1988. Demographic models of the northern spotted owl (*Strix occidentalis caurina*). *Oecologia* 75: 601–7.
67 Cumming 1989.
68 Lee, J.A. 1988. Seals, wolves and words: loaded language in environmental controversy. *Alternatives* 15: 20–9.
69 Martin and de Meulenaer 1988.
70 See, for example, Sinclair, A.R.E. 1983. Management of conservation areas as ecological baseline controls. In *Management of Large Mammals in African Conservation Areas,* R.N. Owen-Smith, ed. Pretoria: Haum.
71 Anderson, J.L. 1983. Sport hunting in national parks: sacrilege or salvation? In *Management of Large Mammals in African Conservation Areas,* R.N. Owen-Smith, ed. Pretoria: Haum.
72 Anderson 1983.
73 Baskin, Y. 1991. Rhino biology: keeping tabs on an endangered species. *Science* 252: 1256–7.
74 Martin, R.B., Taylor, R.D. 1983. Wildlife conservation in a regional land-use context: the Sebungwe region of Zimbabwe. In *Management of Large Mammals in African Conservation Areas*, R.N. Owen-Smith, ed. Pretoria: Haum.
75 Roberts, L. 1991. Ranking the rain forests. *Science* 251: 1559–60.
76 May 1990.
77 See, for example, White, L. 1967. The historical roots of our ecological crisis. *Science* 155: 1203–7.
78 See, for example, Thompson, J. 1991. East Europe's dark dawn. *National Geographic* 179(6): 36–69.
79 Fourie, J., Joubert, S.C., Loader, J.A. 1990. Environmental education: an approach based on the concept of life. *Koedoe* 33: 95–109.
80 Parker, I.S.C. 1983. Conservation, realism and the future. In *Management of Large Mammals in African Conservation Areas*, R.N. Owen-Smith, ed. Pretoria: Haum.

6 The Interface of Health, Population, and Development: The Ecology of Health

REX FENDALL

Almost half a century ago observations were made in the West Frontier districts of India that led to the formation of the Pioneer Health Centre in London in 1926 and, eventually, to the founding of the Peckham Health Centre in London in 1935 for the study of families living within a community to determine those factors that significantly influence health. This study became influential in reorienting medical thinking to the concept of "positive health." The instigators of the Peckham experiment had two main concerns: the fitness of parents for their biological task, and the fitness of the child's home and community environment during the critical first decade of his or her life.

It is to Robert McCarrison, District Medical Officer to the remote Hunza people in the North West Frontier district of India from 1901 to 1911 that credit lies for the original observations. He became aware of the excellent health and longevity enjoyed by the Hunza people despite the harshness of their environment and contrasted this with the squalor and ill-health of neighbouring communities living in similarly harsh conditions.

Health may be defined as a state of harmony between generic man and his environment – physical, social, cultural, and spiritual. Barlow[1] observes that "human beings grow into the world about them by virtue of the systems which form themselves, and correlate and integrate within them. The systems set up the whole, but the whole accommodates the systems." Winslow[2] cites the official records of the World Health Organization (WHO) (1949): "Public health officers have long affirmed that economic development and public health are inseparable and complementary, and that the social, cultural and economic development of a community and its state of health are interdependent." The field of public health is inextricably linked with change. It is a reform movement directed toward the improvement of the quality of life of human beings. The bane of such improvement is the over-rapid growth of populations, which offsets the gains made by technical and scientific discoveries. Scientific and technological changes are also the very means that have created the wealth that could and should facilitate improvements in the living standards of disadvantaged peoples living in deprived circumstances. Standards of living are contingent upon many factors, not all of which are generally considered to be properly within the sphere of direct health action. These factors, as originally formulated by an expert committee of the United Nations in 1954, are:

- health, including demographic conditions
- food and nutrition
- education, including literacy and skills
- conditions of work
- employment situations
- aggregate consumption and savings
- transportation
- housing, including household facilities
- clothing
- recreation and entertainment
- social security
- human freedoms.[3]

If then each of these components contributes to the quality of life as a whole, they are all interdependent one to the other.

DEMOGRAPHIC TRENDS

The most frightening aspect of demographic trends has been until recently the accelerating rate of growth of the world population, which from a low of 0.04 per cent per year around 1800 peaked at around 2 per cent in 1970.[4]

The "doubling time" for world populations has decreased dramatically. It is this shortening of the doubling time that creates problems in endeavours to improve the quality of life, especially for those who are disadvantaged and deprived. The problem is exacerbated by the disparity in growth rates between the richer and poorer countries. This disparity is becoming greater, even though overall growth rates have peaked and are beginning to decline. Total numbers of people are still increasing, and "zero growth" has yet to be achieved. Meanwhile, the poor and deprived increase disproportionately to the "well off."

POVERTY

The world population in 1990 was 5.3 billion. Projections anticipate growth to between 10 and 15 billion by the year 2100 – that is, a doubling or trebling of the 1990 figure, depending upon which forecasts one accepts. The indeterminate factor is not so much mortality rates as uncertainty regarding fertility rate trends.

The population of industrialized countries is 1.2 billion and of the less developed countries 4.1 billion: a 20:80 ratio. If one-fifth of the world cannot successfully succour four-fifths of the world today, what hope is there for the future? In another thirty or forty years the industrialized countries of today will represent only 10 per cent or less of the world population.

Even with a successful gross national product (GNP) growth in developing countries of 4 to 7 per cent (average 4.3 per cent) in the past decade, population growth rates ensure that per capita incomes take a generation or more to double (at constant prices). This fact, added to the very low per capita incomes in developing countries and the very slow effect of the filtering-down process of wealth, does not lead me to believe that poverty in developing countries will be relieved to any substantial degree. There is still considerable poverty in the *developed* regions of the world: witness the slums of the inner cities.

Even though the industrialized countries have a lower GNP growth rate (3 per cent for the past decade) their per capita GNP will be greater because of the very low population growth rate. Therefore the disparity in per capita incomes will increase, not decrease (Tables 6.1 and 6.2).

Michael Walton[5] affirms that "the conditions of the poor are grim." He presents a table projecting that in the year 2000 poverty, which he defines as "the inability to attain a minimal standard of living," will afflict 18 per cent of the population of developing regions, or 825 million people.

Absolute poverty is a condition of life so characterized by malnutrition, illiteracy, disease, squalid surroundings, high infant mortality rate (IMR) and low life expectancy as to be beneath any reasonable definition of human decency. This definition does not include less extreme but still severe and unacceptable degrees of poverty.

Table 6.1
Per Capita GNP Growth in $U.S.

	1981	*1990*	*Factor*
Industrial countries	7,260	15,830	2.18
Less developed countries	560	710	1.27
Latin America and Caribbean	1,580	1,930	1.22
Sub-Saharan Africa			
Incl. South Africa	512	465	0.91
Ex. South Africa	411	305	0.74
India	190	330	1.70
China	230	330	1.43

Data source: World Population Data Sheets 1990

Table 6.2
GDP and Population Growth

	% growth GDP 1980–89	*% population growth 1990*	*% per capita growth 1980–89*
Industrial countries	3.0	0.5	2.5
Less developed countries	4.3	2.1	2.3
Sub-Saharan Africa	1.0	3.0	–2.0
Latin America and Caribbean	1.6	2.1	–0.6
India	5.6	2.1	3.5
China	10.1	1.4	8.7

Data sources: World Population Data Sheet 1990; World Bank 1990

Rao and Sastry[6] suggest that the traditional statistics used to measure improvement in living standards do not reliably reflect change. They suggest that measuring growth retardation on a weight-for-age index would give a more reliable indication of "progress" for programs of health intervention, nutrition education, and income generation. A better measure of poverty levels, perhaps?

There is little doubt in my mind that the number of people affected by poverty has risen consistently over the years. Whereas *proportions* of deprived persons may have decreased, *actual numbers* have risen, not only in the huge visible urban conglomerations, but also in the villages and small towns of rural areas, and certainly in the peri-urban septic fringes of cities in less developed countries.

Let me quote from Michael Camdessus's statement of the aims of the International Monetary Fund (IMF) in helping developing countries:[7]

"Our prime objective," he writes, "is [economic] growth." Growth, he avers, means "high quality growth" – not "pseudo growth," not "flash in the pan growth," not "growth for the privileged few leaving the poor with nothing but empty promises," and "not growth through disorderly exploitation of natural resources and the ravaging of the environment." High-quality growth, Camdessus states, means "sustainable and dynamic [growth] that invests in *human* capital [and] is concerned with the poor, the weak and the vulnerable. It is growth that does not wreak havoc with the atmosphere, with the rivers, forests or oceans." (Can anyone tell me where this happens?) Camdessus continues: "high quality growth will not be achieved in the absence of bold reformism." I agree, and would add that reformism in economic development will not succeed unless runaway and wasteful population trends are restrained. In its 1990 report on world population the United Nations Fund for Population Activities (UNFPA) stated that "absolute poverty has shown a dogged tendency to rise in numerical terms. The poorest 20 per cent of the population still dispose of only 4 per cent of the world's wealth."[8]

Poverty is one of the fundamental causes – along with pestilence, squalor, and ignorance – of ill-health. Economic growth is inextricably linked with population growth. There has not been in my lifetime any dramatic improvement in the situation of the deprived, either quantitatively or qualitatively.

HUNGER

The population density of the world doubled from about 12 persons per square kilometre in 1900 to 22 by the mid-sixties and is estimated to double again by the turn of the century. But whereas the developed world will have doubled its population density from approximately 10 persons per square kilometre in 1900 to 22 by the year 2000, in the less developed nations this will have increased by a factor of five over the same period.

The UNFPA report states: "The incidence of malnutrition declined from 27% of people in developing countries in 1969–1971 to 21.5% in 1983–1985. But because of population growth the total number of malnourished increased from 460 millions to 512 millions, and is projected to increase further to 532 millions or more by the end of the century."[9] This will occur, notwithstanding the fact that average world daily caloric intakes rose by 20 per cent (from 2100 to 2500) between 1965 and 1985. There is still a gross disparity between the rich and poor nations: in developed countries the average diet provides approximately 3500 calories daily, as compared with 2500 calories daily in less developed countries. Moreover, the latter figure is reached only by a growing reliance on imports from North America. This is not a healthy situation.

By far the greatest number of the malnourished are children under five suffering from marasmus and kwashiorkor. Whether by direct cause, starvation, or through succumbing to a concomitant infection, these unfortunate children account for a substantial proportion of the excessive morbidity and mortality in their societies. Mortality rates of up to one in three of the under-fives is not tolerable in any civilized society. More children in developing countries die of energy–protein malnutrition than from any other specific disease. Much of the tropical world is worse fed now than it was before World War II. Periodic famines, droughts, floods, pestilence (locusts), and wars add acute crises to a chronic situation and seem to be more frequent than hitherto. Nutrition and infection are well noted to be interrelated. The mortality rate of preschool children has been overwhelmingly less responsive to change than has been the infant mortality rate, and in my opinion is a more sensitive index of progress.

The number of members in a family is a decisive factor in the nutritional status of that family. It has been observed (in Ankara) that in the slum and squatter areas of towns where families are large one-third of such families show signs of calorie deficiency, and that if each of these families had had one child fewer, the proportion of calorie-deficient families would fall to 10 per cent.

Four basic factors contribute to poor nutritional status: (1) imbalance between food production, supply, and population growth; (2) excessive fertility; (3) ignorance and illiteracy; and (4) inadequate income. The outlook, according to the UNFPA report for 1990, is gloomy: developing countries, "as a whole, have suffered a serious decline in food self-sufficiency."[10] The report states that although theoretically the world could produce sufficient food for 14 billion persons, in reality the situation in developing countries in getting worse: thirty-six countries with nearly half a billion people will likely be unable to feed their own populations from their own lands by the year 2000.[11] Where there are deficiencies it is the young, the old, and women who suffer disproportionately. Female children tend to suffer a greater incidence of kwashiorkor than male children. The age of menarche tends to be higher among undernourished girls, and pregnancy wastage, premature delivery, maternal mortality, and stillbirths are more common. Lactation and fertility are adversely affected by malnutrition. Prolonged breastfeeding by women on marginal diets leads to further nutritional stress and deficiencies and low body weight. It should also be borne in mind that whereas the under-fives account for approximately 10 per cent of the population of industrial countries and only 5 per cent of deaths, in developing countries they account for nearly 17 per cent of the population and one-third of the deaths, and that mortality rates among the very young have a significant influence upon cultural attitudes to family size and family limitation.

URBANIZATION AND POPULATION DISTRIBUTION

Urbanization is essentially a twentieth-century phenomenon (Table 6.3). The industrialized nations, with their slower urban growth and wealthier economies, have still thrown up inner-city slums where diseases associated with overcrowding, poor sanitation, and squalor are common. This is not to mention the growing phenomenon of cardboard cities. The towns of developing countries have thrown up peri-urban shanty towns, in addition to inner city slums, so that their deprived populations range from 10 to 60 per cent of the total. These people exist in a state of total deprivation, enduring the hardship of absolute poverty, squalor, deficient sanitation, and inadequate water supply in improvised and overcrowded housing.

It is not urbanization in itself that is at the root of the problem; rather, it is the over-rapid growth of urbanization that has outstripped economic, social, and management resources. Urbanization, a monetary economy, industrialization, and education are generally considered to be factors contributing to lower fertility trends, but I am not aware that this applies to *deprived* urban families. For example, in Sri Lanka the crude birth rate in the mid-seventies for Colombo City was 20.4, but the rate for the urban poor, the slum and shanty-town dwellers, has been quoted at 28.0.[12]

Table 6.3
Percentage Urban Population

Region	*Mid-60s*	*1990*	*Predicted urban growth rate 1990–95*
U.K.	80	92	0.3
U.S.A.	60	74	0.8
Latin America	30	72	2.7
Asia	20	29	3.3
Africa	15	34	4.9

Data source: UNFPA 1990

These peri-urban shanty towns, or squatter towns, form a septic fringe around the towns proper and constitute, with their migrating populations of casual workers, a focus of disease – social, mental, and physical – that spreads inward to the metropolitan areas and outward to rural areas.

While urban populations in the more developed countries are expected to expand by some 65 per cent between 1970 and 2000, that increase in the less developed countries is likely to be nearly 250 per cent. This is an extremely rapid urban growth rate, and represents perhaps the most disruptive social factor. Standards of living for a fast-growing sector of urban dwellers appear to be getting worse, not better. Urban economies from New York to Calcutta cannot keep pace with growing demands. Population growth by far outstrips economic growth and leads to ever-increasing numbers of people trapped in squalor, poverty, ill-health, and malnutrition.

RURAL ASPECTS

Notwithstanding the growth of urban populations and the diminishing proportion of rural inhabitants in developing countries, the actual numbers of people living in rural areas continues to increase in developing countries. Thus, developing countries will also have to cope with growing rural populations while industrialized countries adjust to a decrease in rural populations. With increasing rural densities, fragmentation of landholdings, increasing underemployment and unemployment, disruption of village culture, as well as a disadvantaged infrastructure of roads, transport, schools, and health services, the living conditions and health status of people living in developing countries are likely to worsen. This will give rise to problems in the provision of health care, particularly where population density is very low and communications totally inadequate. In some situations, a pastoral way of life that necessitates internal seasonal migration will aggravate these problems still further.

It appears to me that we should not separate the urban and rural health issues that result primarily from the drift to the towns (as a consequence of increasing rural populations, decreasingly available arable land and degradation of such land) since disease and ill-health from the peri-urban septic fringes of towns spread to both rural and true urban communities. These issues are interrelated; solutions for one group must not be implemented at the expense of the other.

The unequal adoption of family planning, for example, between rural and urban populations has important consequences for both sectors. In China, only 8 per cent of rural women comply with the state's one-child policy, while 61 per cent of urban women do so.[13] In the long run this may create real social effects, such as a shortage of young people in the cities to care for elderly parents. In other countries it is the migration of young people to the cities that leads to the same problem. The balance of sexes can also be disrupted, depending on whether it is mostly boys or girls who migrate to the cities.

THE HOME AND FAMILY ENVIRONMENT

The Public Health Reform Movement in England commenced effectively in 1848 with the first of the great Public Health Acts. Informed by a new understanding of the epidemiology of some of the major scourges of the period, the Act concerned itself primarily with sanitary engineering, i.e. the provision of adequate housing, pure water supplies, and safe sewerage systems. It is possible to consider the evolution of the Public Health Reform Movement in terms of four phases. In the first, lasting from 1848 to 1870, problems of sanitation were addressed. This initiated the decline of common alimentary and respiratory diseases. In the second, scientific, phase (lasting from 1870 to 1920) gains were made with respect to infectious diseases, malnutrition, and growth retardation thanks to the discovery of vaccines and the development of technology (Figures 6.1–3). The third phase (1910–1940) was characterized by a concern with personal health services, including health insurance (e.g. the Lloyd George Penny Insurance Scheme) as well as with disease prevention and health promotion services, partly as a response to the findings of health examinations for Boer War recruits. These concerns were dominant until the end of the Second World War, when (in 1948) the era of socialized medicine under the National Health Service was ushered in.

In stark contrast to the improvements in public health that have been made in Britain and other industrialized nations, throughout the developing countries housing, sanitation, and water supplies (the basic requirements for improving health) remain deplorably deficient in both quality and quantity in urban and rural areas alike. Vast improvements in physical amenities are needed, and, along with these, extensive educational programs to ensure the proper understanding and utilization of these amenities.

A home should offer amenities for daily living as well as adequate living space inside and outside the dwelling. The inadequacy of housing in the developing world is painfully obvious: the number of urban "street children" is estimated at 80 million.[14] Housing shortages in urban areas are relatively easy to assess, as they are reflected in the

Figure 6.1
Mortality from enteric fever, England and Wales 1871–1971. *Data source*: DHSS 1976 *Prevention and Health: Everybody's Business*. London: HMSO.

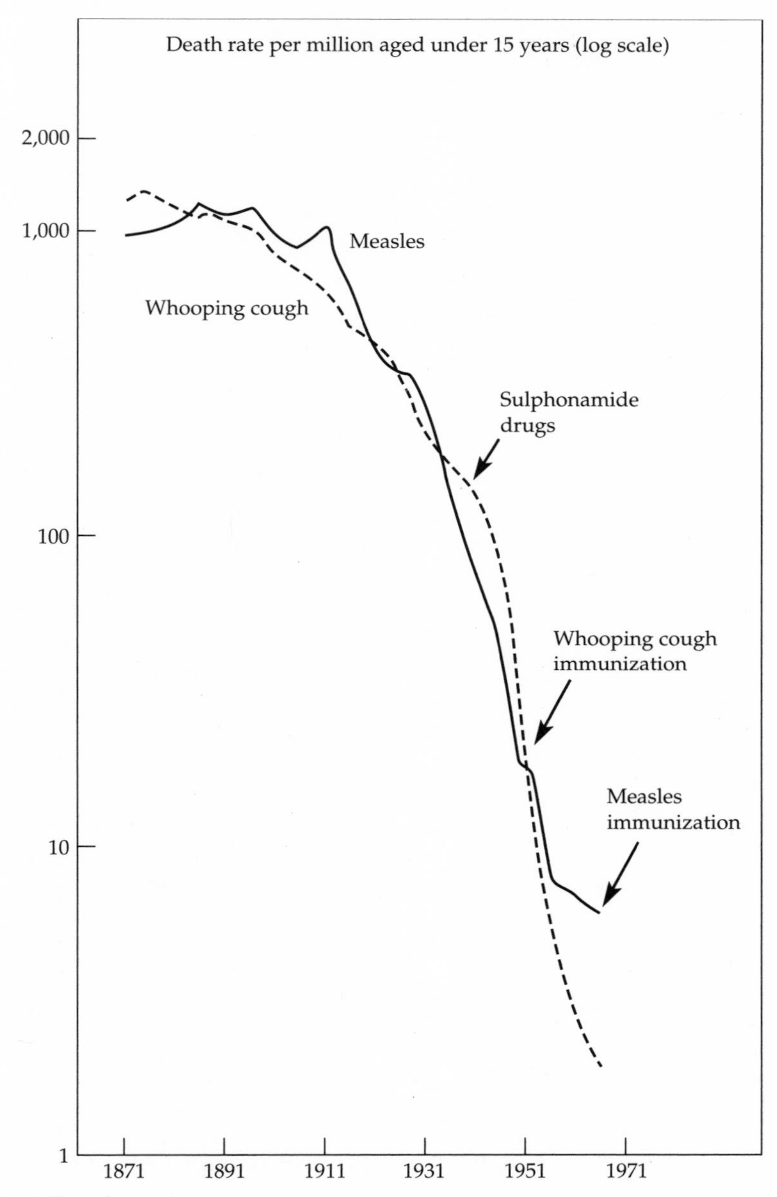

Figure 6.2
Childhood mortality from measles and whooping cough, England and Wales, 1871–1971. *Data source*: DHSS 1976 *Prevention and Health: Everybody's Business*. London: HMSO.

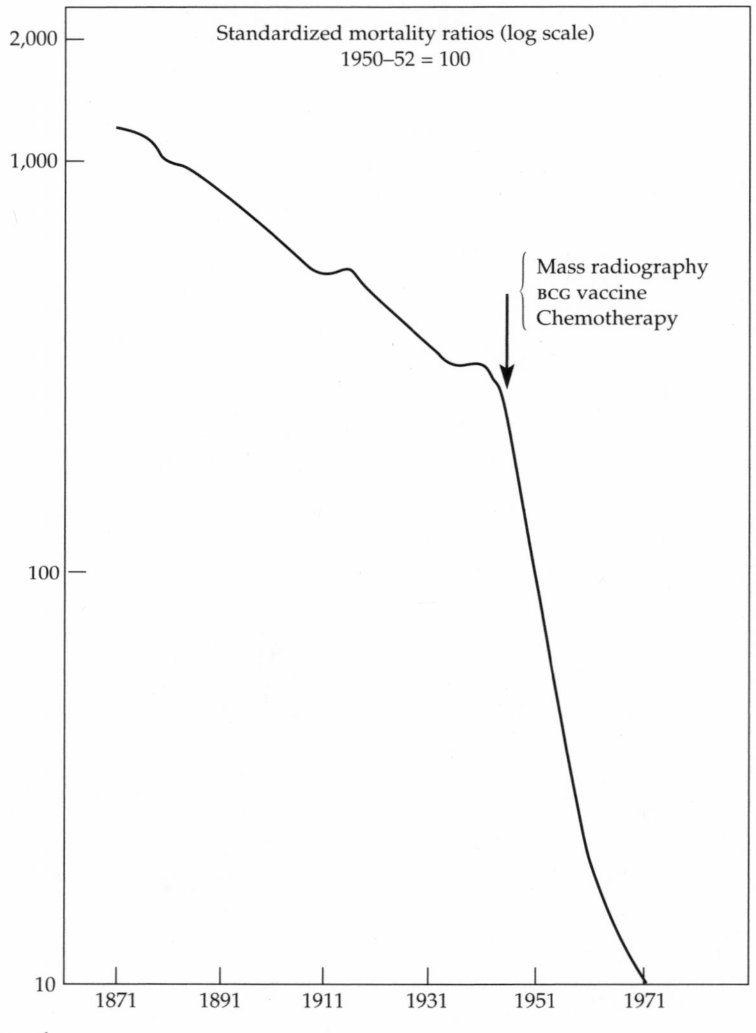

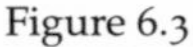

Figure 6.3
Mortality from tuberculosis, England and Wales, 1871–1971. *Data source*: DHSS 1976 *Prevention and Health: Everybody's Business*. London: HMSO.

proportion of town populations living, or subsisting, in shanty towns: for example, 10 per cent in Kingston, Jamaica, 30 per cent in Karachi, and 45 per cent in Ankara. According to the UNFPA report for 1990, "of every 100 new households in developing countries 72 live in shanties or slums, 92 out of every 100 in Africa."[15] In rural areas the problem of deficient housing is also extensive, although it is more difficult

	Replacement requirements	Increase of population	Total
Latin America	26%	74%	30 million delta
Africa	27%	73%	34.2 million delta
Asia	27%	73%	106.7 million delta

Figure 6.4
Estimated housing requirements 1970–80. *Data source*: U.N. 1974 *World Housing Survey.*
Note: "Delta" is defined as the useful housing solution required to accommodate a household under variable standards (a) to replace existing stock at a rate of 2 per cent per annum (b) assuming an occupancy rate of 1.5 household per delta.

to measure. Overall, the deficiency is estimated at about half a billion dwellings. Of this total, one-fourth can be attributed to shortages, one-fourth to one-third to obsolescence, and one-third to one-half to the increase in demand made by growing populations (Figure 6.4).

Moreover, even when new dwelling schemes are developed there is no guarantee that they will benefit the poor. Malacoo states that 60 per cent of the urban housing program in Kenya in 1980 was directed towards upgrading squatter areas by site and service schemes. He observes that the "allocation of plots was susceptible to 'downward raiding' by the upper income groups and that many of the plots change hands from the original allottee."[16] In a report on the Dandra Community Development Project in Nairobi, Lee-Smith and Menon comment that while implementation was successful in terms of self-help building, it "has not been possible to confine ownership of the project plots to the city's low income population … However, there was attempted and actual interference of plot reallocation."[17]

There are many environmental programs that, by the time they are completed, leave more people *unserviced* than there were at the beginning, despite an apparently favourable increase in percentages. For example, Wolman and Hollis,[18] writing on a generation of progress in sanitary services of water and sewerage schemes in Latin America between 1961 and 1969 (covering some 200 to 300 million people), demonstrate that water supplies coverage increased from 33 to 46 per

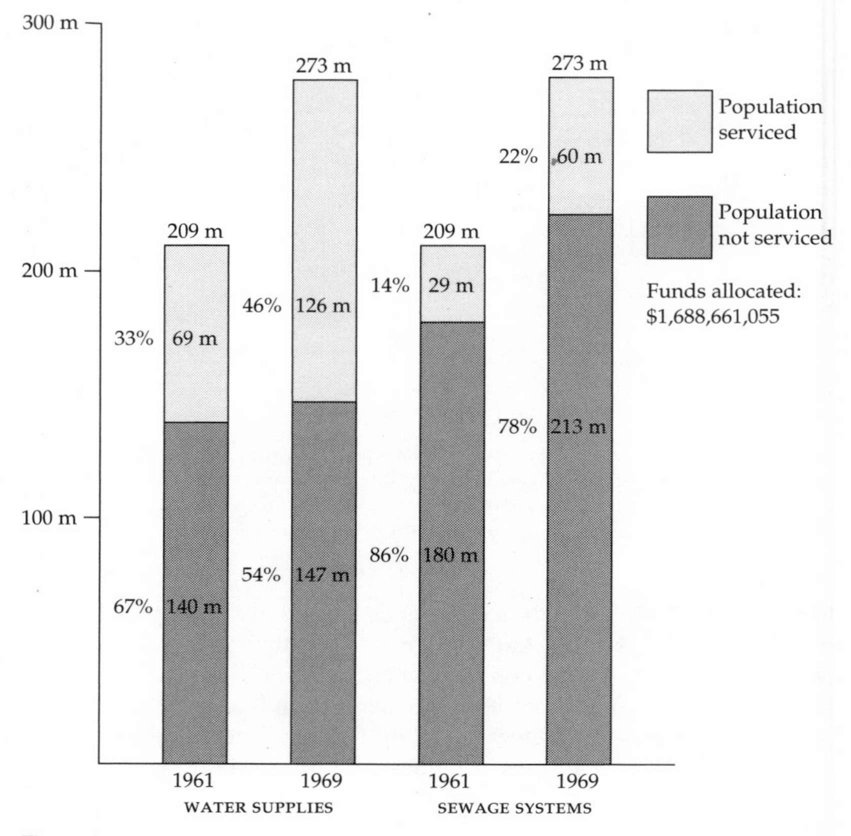

Figure 6.5
Growth of water supply and sewerage systems, Latin America and the Caribbean.
Data source: Wolman and Hollis 1971.

cent of the population and sewerage from 14 to 22 per cent. But the numbers of people *unserviced* increased from 140 to 147 million and 180 to 213 million respectively, simply because the population had increased from 209 to 273 million over that period. The cost was U.S.$1.7 billion (Figure 6.5).

I need not stress the importance of safe water and safe sanitation in reducing diseases and improving personal and public hygiene. In 1976, the United Nations Conference on Human Settlements (Habitat) set 1990 as the target date for a total outreach of safe water and safe sanitation to both urban and rural peoples.[19] In 1980 the UN launched the "Water Decade." Writing towards the end of that decade, Rotival,[20] citing the World Bank, states that water coverage "in cities will increase from 35% in 1985 to 40% by 1990" and "rural water supplies from 32% in 1981 to not exceeding 45% in 1990." Overall, he writes, "the supply of potable water to urban and rural populations is unlikely to exceed 68% coverage" in 1990; that is, one third of the world population will *not* have potable, let alone safe, water. Quoting the WHO, he writes: "In respect to developing countries less than 50% have access to adequate water supplies and only 20% have adequate sanitation facilities." The WHO Review of Primary Health Care Development[21] states that sanitation coverage decreased from 33 per cent to 25 per cent in the 1975–80 period. It ascribes this situation to the population increase, rural–urban migration, and economic stagnation in developing countries.

How, then, are we to overcome the water-borne diseases? The World Bank estimates that U.S.$9 billion is required annually, utilizing low-cost, sustainable, and replicable technologies. Not only are we a long way from reaching that target but even water itself is becoming a scarce commodity. The "green revolution" utilizes much more water for crops; this, in addition to increasing industrial use, population growth, and rising per capita consumption ensures that targets for piped water supplies and water-borne sanitation elude us more and more. The UNFPA reports that the total number of people without safe sanitation increased from 1,034 million to 1,750 million between 1970 and 1985; in other words 30 per cent of the world's population was without adequate sanitation at that time.[22] This figure probably underestimates the current problem. The results of the housing census in Sri Lanka in 1981 indicated overall that while 31 per cent had no toilet facilities another 38 per cent had only pit latrines, and that only 15 per cent of *urban* housing had flush toilets. As Gunatilleke and colleagues state, "The Sri Lankan performance in the health sector has been widely recognized as being exceptional for a country at a (then) level of per capita income in the region of US $300" (1990 $420); average economic growth was about 5% to 6 % during the 1980s and the population growth rate 1.6 per cent, with a 75 per cent rural population.[23]

In Sri Lanka, piped water supplies are cited by Gunalilleke and colleagues as reaching 45 per cent of urban populations, 5 per cent of rural populations, and 75 per cent of estate dwellers: populations serviced are stated as 2 million, 0.4 million, and 0.8 million respectively. But 80 per cent of the water supplies to the urban population are said to be inadequate and unsafe. The housing situation described in this Sri Lanka study is equally appalling, with an average living space of 100 square feet per person. Slum and shanty dwellers form 43 per cent of Colombo's population and occupy only 2 per cent of

Colombo's 170 acres. Housing ranges from modern dwellings with luxury fittings, to shanties without water and toilet facilities, and overcrowded housing units in slums with one standpipe for 257 persons, one toilet per 36 persons, broken foul-smelling drains, and pits on all sides. The study puts it mildly: "The shanty dwellers are exposed to grave inconveniences in their housing environment, which lacks basic facilities and services."[24]

But progress *is* being made in that the "water decade" has become a "water supply and sanitation program" and is accepted as part of the primary health care movement. Research has concentrated upon practical issues such as the relationship between disease and poor water supply and the concept of "appropriate" technology, thus recognizing the importance of village and rural water supply and sanitation systems as well as of high-technology water supply and sewerage systems.

A shortage of appropriately trained professional and auxiliary personnel for housing, water supplies, sanitation, sewerage, and waste disposal continues to hamper efforts to overcome the backlog of needs and keep pace with urban and rural population growth. A community approach that encourages local participation in projects will in itself demand a dramatic increase in ground-level educational programs if lack of maintenance and improper utilization are not to frustrate the objectives of the Habitat schemes.[25]

The dilemma is to choose between economic and socioeconomic development, the latter being a much more expensive prospect. In effect, a healthy home environment does not exist for the majority of urban and rural families in developing countries, but the very worst conditions are those of the "septic fringe" of the cities of developing countries where living conditions are worse than those of the inner-city slums of industrialized countries.

To sum up, the majority of people in developing countries are suffering the hardship of insufficient, inadequate, and obsolete houses, deprivation of living space both inside and outside the "hovel," inadequate and unsafe water supplies, sewerage conditions that beggar description, an absence of lighting and ventilation, lack of privacy, and overcrowding; urban dwellers are beset by the additional problems of air pollution, solid waste, excessive noise, high accident rates, social disintegration, crime, and prostitution. Underemployment and fragmentation of the land in rural areas gives way to frank unemployment in the towns. The rate of population growth exceeds the capacity of economic growth to reduce the deprivation of the majority of humanity. Given the differential growth rate in both population and per capita incomes, that deprived segment grows ever larger.

MATERNAL AND CHILD HEALTH

Population growth results from a differential between birth rates and death rates; the higher the latter, the greater the fertility rate. This effectively creates a high wastage rate and has an adverse effect on maternal health. The overall fertility rate in the less developed nations, where, on average, 4.2 children are born to each woman, is twice that of the more developed countries. Sub-Saharan Africa has an average rate of between 6 and 7 whilst Rwanda achieves a rate of 8.3. High parity leads to maternal exhaustion, and since maternal mortality increases with advancing age and with increasing parity, there appears to be a correlation between multiparity and obstetric and gynaecological complications. Maternal illnesses in general increase with family size and short birth intervals. Large families in developing countries also tend to have a history of several abortions, whether spontaneous, legal, or illegal. A woman may in her lifetime have twelve or more pregnancies to achieve a family size of six or seven children and ensure one surviving adult son. Reporting on a study of maternal mortality in India, Bhatia states that the number of deaths revealed in the study was much higher (8/1000 live births) than official statistics revealed, and that many women died because of inadequate transport facilities.[26] Maternal deaths accounted for one-third of female mortality in the reproductive age group. The rates varied substantially from place to place, reflecting differing levels of economic development and the presence or absence of primary health centres and subcentres.

Child health is also affected by excessive population growth rates. The larger the family size the less individual attention each child receives. The proportion of children in the Third World population is that much higher: the percentage of those under 15 years being 40 per cent as opposed to 22 per cent in industrialized countries. Malnutrition is more likely to be evident in larger and poorer families. There is a definite correlation of fetal and infant mortality with pregnancy order. Apart from the first, where both are high, they rise with each succeeding pregnancy, particularly with the higher orders of pregnancy. There is a higher perinatal mortality rate with younger and older primiparae, and older multiparae. The highest perinatal mortality is associated with the highest parity; it is lowest with the second birth and increases dramatically with the fifth and subsequent births. Increased birth spacing gives each child a better chance of survival, and better maternal care due to less maternal exhaustion. These factors, together with disease and poverty, ensure that infant mortality rates, though falling, continue to be vastly excessive in the developing countries: 91/1000 live births as compared to 16/1000 in the industrialized countries. In most countries in sub-Saharan Africa, infant mortality rates exceed 100 – e.g., 154 in Ethiopia, 110 in Pakistan, 166 in East Timor – whereas in Europe the average is 10 and, in North America, 7 (5.9 in Finland, 9.5 in the U.K., and 7.3 in Canada). Furthermore, if these differences are unacceptable, the disparity between the death rates of the under-twos and the under-fives are even more so. One out of three or one out of two children born in developing countries will die before reaching his or her fifth birthday. (Undernutrition and malnutrition are important factors in this morbidity/mortality pattern, and probably affect up to one in four or five families in developing countries, being more prevalent among preschool children and lactating mothers.) Another way to quantify the problem is to point out that whereas deaths of children under five represent roughly half their proportion in the population in well-developed countries, in the less well-developed nations they account for at least twice the proportion of under-fives. This is an intolerable waste, even in the practical terms of human endeavour, health care efforts, food production, national economic resources, and world resources.

As the *WHO Review of Primary Health Care Development* states, "The complex interrelationships between infant and child mortality are being matched by the complex relationships between population policies and health policies."[27] Again, referring to the increasing complexity of the role of women in the health and development of the family and community it states, "The link is recognised between the level of maternal education and child mortality."[28]

EDUCATION AND HEALTH CARE

If we believe that literacy and learning are essential to the improvement of public health, then we must be dismayed at recent illiteracy figures that show an absolute increase from 742 million to 889 million persons despite a decrease in the proportion of illiterate people from 32 to 28 per cent between 1970 and 1985. Though the number of primary school children enrolled has risen from 395 to 665 million, and secondary school enrolment has more than doubled from 79 million to 175 million, the total number of children out of school rose from 284 million in 1970 to 293 million by 1985, with a projected increase to 315 million by the year 2000.[29] Moreover, higher enrolments at primary and secondary schools do not guarantee an equivalent increase in graduates, and high "drop-out" rates create the additional problems of unemployability and discontent among youth. All this

Table 6.4
Distribution of Medical Schools Worldwide 1970/71

Region	*Population (× 10^6)*	*Medical schools*	*Population per school (× 10^6)*
Latin America	291	157	1.8
Africa	354	41	8.6
Asia (including China)[1]	2,104	282	7.5
Developing nations	2,749	480	5.7
North America	229	123	1.8
Europe	466	219	2.1
U.S.S.R.	245	83	3.0
Oceania	20	12[2]	1.7
Developed nations	960	437	2.2
WORLD	3,709	917	4.0

[1] China population 773 million, 41 schools; Japan (industrial) population 105 million, 46 schools
[2] Includes 1 Fiji, 1 Papua

Data source: World Directory of Medical Schools. Geneva: WHO

is attributable to an imbalance between population growth and economic growth whereby an adequate number of school places cannot be provided for eligible children. There is also still a serious imbalance between the percentage of males and females enrolling for secondary schooling, a factor that must have an adverse effect on health education of mothers and mothers-to-be with regard to prenatal and maternity care and child development. Out of 44 countries in Africa there is not one country in which female secondary education enrolment exceeds that of males; in Asia, only 6 out of 30 countries provide figures on this question, and in Latin America and South America 12 out of 18 countries provide figures. Of the 44 nations in Africa, only Egypt has a female secondary school enrolment exceeding 50 per cent of those eligible; in Asia (excluding Eastern Asia) only Singapore, Malaysia, Sri Lanka, and five countries of Western Asia report this level. It is against this background that we must consider the technical education for health care personnel.

With regard to the education of physicians over the years 1970/71 to 1982/83, although the number of medical schools worldwide increased from 917 to 1,165, the ratio of medical schools to population has remained constant at one school per 4 million inhabitants (Tables 6.4 and 6.5). In industrialized nations population to medical school ratios have decreased from one school per 2.2 million to one

Table 6.5
Distribution of Medical Schools Worldwide 1982/83

Region	*Population* (× 10^6)	*Medical schools*	*Population per school* (× 10^6)
Latin America	381	204	1.868
Africa	498	73	7.014
Asia (including China 54)	2,739	362	7.608
Developing nations	3,618	639	5.698
North America	252	145	1.738
Europe	421	275	1.531
U.S.S.R.	269	92	2.924
Oceania	19	14	1.357
Developed nations	961	526	1.827
WORLD	4,579	1,165	3.944

Data sources: *Directory of Medical Schools Worldwide* 1982. Geneva: WHO; Association of Commonwealth Universities 1983; *World of Learning* 1982/83; *World Almanac* 1983

Table compiled by Julian Kyne, N.R.E. Fendall, and K. Ofosu-Barko

per 1.8 million (an increase of 96 schools). In the developing nations, despite an increase in the number of medical schools from 480 to 639, the ratio has remained constant at one medical school per 5.7 million people, a standstill growth rate that has profound implications for health manpower staffing plans based on the primacy of the physician in delivering health care. This is particularly the case if we take into consideration the proportion of specialist physicians, the distribution between primary care and referral care, the distribution between town and country, placement in private and public sectors, and the emigration factor from poorer to richer countries. Emigration can be a costly factor in manpower production within professional cadres, with regard to both education and subsequent employment. Figures quoted by the UNFPA[30] vary from two-thirds of doctors emigrating from Thailand to between one-quarter and one-fifth emigrating from countries such as India, the Philippines, Central America, and Turkey (of the numbers of physicians graduating annually).

The same issues arise in relation to other health educational institutes and professional health personnel. Can the poorer countries keep pace with, still less outstrip, population growth rate to provide an acceptable increase in density and quality of professional manpower? Clearly not. The growth of institutions, facilities, and health manpower is insufficient to make a rapid and meaningful impact in developing countries. In the next forty years the population of the poorer countries is expected to double to 8 billion. In that time facilities and manpower would also need to double in order simply to maintain present levels of service, and to double again to increase the penetration of health services so that existing ratios of facilities and manpower to population are improved. In other words, a quadrupling of existing health manpower, hospitals and hospital beds, drug supplies, and facilities will be required. Teaching institutes and teacher training will need to expand at comparable rates, and at the same time standards must be raised. Is this feasible? Would it be regarded as adequate progress over two generations, given the rapidly rising expectations of communities? Even if the number of medical, midwifery, nursing, and pharmaceutical schools could be increased sufficiently to ensure an adequate penetration of both rural and urban populations, would the education they provide be relevant to actual needs? The teaching of doctors and nurses has been heavily oriented towards hospital and specialist care and high technology. Meanwhile, health services in developing countries are rapidly changing their orientation in favour of primary health care.

Despite the efforts of the community-orientated Educational Institute for Health Sciences to shift the emphasis from institutional and referral care medicine to community and primary care, the transition of curricula overall remains painfully slow and inadequate. There is, as I have said elsewhere, "a basic mismatch between the goals of health care for all and the goals of medical education."[31] A plea for more collaboration within medical schools of the Commonwealth by Marwah and myself[32] ended by proposing "the concept of a Commonwealth University devoted solely towards the needs of development." This has now come to pass, but with what effect we shall have to wait and see!

HEALTH MANPOWER

Health manpower statistics do not provide a clear or reliable basis for examining and predicting change. The technical notes to the World Bank's *World Development Report* for 1990 urge the reader to interpret its findings cautiously, as many data are "subject to considerable margins of error" and "inter-country and inter-temporal comparisons always involve complex technical problems."[33]

Comparisons of numbers and ratios of health personnel to population are very crude and all befogged by factors such as specialization, rural–urban distribution, migration (both in and out), foreign personnel, and inaccuracies in registration. (While numbers of new graduates are entered punctiliously, the same cannot be said for withdrawals due to emigration, retirement or death.) Comparisons are also affected by a certain vagueness in attention to procedures, in defining categories of personnel, and in making alterations to standards of qualification. Quite a few medical schools began by training personnel to sub-professional standards, e.g., the medical assistants of Madras, Kampala, and Khartoum and the sub-assistant surgeons of several Indian colleges. Similarly, in some countries auxiliary nurses and medical assistants were incorporated in nursing registers and included in reporting manpower figures.

As an example I quote figures from two "reliable" sources[34] concerning the population per physician over the years in Sri Lanka. These figures are as follows: in 1965, 5,820 people per physician; in 1975, 6,323; in 1980, 7,397; and in 1984, 5,520 (only slightly lower than the 1965 figure). During this period medical schools had increased from 1 to 4, output of graduates had increased, and so had physician emigration. Thus, despite considerable effort and economic expenditure, no significant improvement in physician to population ratios had taken place over twenty years, for several of the reasons outlined earlier.

Other countries such as Jamaica show a slight regression for both physician to population and nurse to population ratios over the same period. The physician–population ratio changed from 1:1990 to 1:2040 and the nurse–population ratio from 1:340 to 1:490. Other countries showed dramatic improvements; in Egypt, the physician–population ratio changed from 1:2300 to 1:770 and the nurse–population ratio from 1:2030 to 1:780 over the same twenty-year period. Regions are reported as having improved their ratios of professional doctors and nurses despite population growth over the intervening twenty years, but the fact remains that without that population growth the ratios would have been substantially better (Table 6.6). These improvements in ratios have taken a generation to accomplish. Progress by generation-steps will take an exceedingly large number of years before the poorer countries of the world achieve Canadian or Swedish health personnel–population ratios. To put the matter another way, Sweden has one physician for every five babies born each year whilst Kenya has one physician for every five hundred babies born annually, and no hope of achieving Sweden's standard of maternal and child care in the foreseeable future if Kenya seeks to emulate the Western pattern of health care services.

The other components of health – employment and working conditions, transportation, clothing, recreation, social security, and freedom

Table 6.6
Health Manpower Growth 1965–84

Region[1]	*Population per physician* 1965	1984	*Population per nurse* 1965	1984	*Approximate population doubling time (years)*
LOW AND MIDDLE INCOME COUNTRIES					
Sub-Saharan Africa	32,200	23,850	5,420	2,460	< 25
East Asia	5,600	2,390	4,050	1,570	70
South Asia	6,220	3,570	8,380	2,710	< 30
Europe, Middle East and North Africa	4,760	2,430	3,440	1,160	> 30
Latin American and the Caribbean	2,370	1,230	2,090	1,020	30
HIGH INCOME COUNTRIES					
Sweden	910	390	310	100	> 350
Canada	770	510	190	120	> 85

[1] Covers the regions of sub-Saharan Africa, East Africa, South Asia, Europe, the Middle East and North Africa, Latin America, and the Caribbean

Data source: World Bank 1990

– have all yet to reach desirable standards and outreach. The land still provides over half of the employment opportunities in the developing countries, with underemployment and landlessness in the rural areas at least equalling frank unemployment in the towns, as growing rural–urban migration attests. Conditions of employment and social security systems are reminiscent of the early nineteenth century in Europe. In rural areas the work burden falls most heavily on the woman, in field and home. Transportation leaves much to be desired, especially for the sick who require access to health services. UNICEF figures on access to health services by percentage of population[35] show that in Africa nineteen countries report failure to achieve more than 50 per cent outreach, and that Latin America and Asia each report ten countries failing to achieve 75 per cent outreach (out of some eighty reporting countries). This represents something like half a billion people without adequate access to health care (however this might be defined). (I suspect from my own experience that this figure would be much higher if time commitments and difficulty of travel were taken into account along with distance.) Clothing has much improved, but lack of easy access to clean water supplies often leads one to wonder whether clothes are advantageous or disadvantageous: witness the extent of skin diseases such as scabies, impetigo, eczema, and so forth, which account for some 4 per cent of ambulant care diagnoses in developing countries. Social security does not exist in rural areas, apart from whatever security might derive from the extended family, and is totally inadequate in urban areas, where the nuclear family prevails.

AGING

While the proportion of the young decreases steadily after an initial expansion, that of the elderly grows steadily from the combined results of this demographic transition and an increasing life expectancy. The WHO estimates that worldwide there will be some 600 million persons aged over sixty years by the turn of the century, of whom two out of three will be in developing countries (in 1960 the ratio was 1:1).[36] Bicknell and Parks, quoting the UN Report of the World Assembly on Aging, state that "by the year 2025 the proportion of elderly in developing countries will be 72%."[37] In terms of numbers alone the aging of the population is an increasingly important issue; in terms of the quality of medical and health care required it is also a more difficult, prolonged, and costly problem than that of child care. Children suffer mainly from acute illnesses and elderly people from chronic conditions. The diseases of the young are more responsive to medical and health interventions; those of the elderly are less responsive and much more prolonged. Childhood morbidity and mortality has its roots largely in poor environmental conditions (housing, water, and sanitation), whereas those of the elderly result predominantly from stress and degeneration. Both groups suffer from malnutrition. Aging has both a disease component – cancer, strokes, heart failure, osteoarthritis, diabetes – and an infirmity component due to physiological changes – impaired hearing and sight, skin sensitivity, a weakened immunity system, osteoporosis, imbalance, incontinence – in a word, frailty, which demands constant care and attention.

Not only is the aged population growing at a faster rate than the overall population but with an ever-increasing life expectancy the mean age of the elderly is rising. The majority of the elderly are neither totally disabled nor totally dependent and do not necessarily require hospital or institutional care. But for those who are infirm and do require support, there is a growing need for primary health care, medical and nursing care, and social support, preferably in the home. But, as Bicknell and Parks point out, "we are not doing a good job for our own populations facing chronic diseases and aging."[38] In the U.K. 50 per cent of all hospital expenditure can be attributed to the 16 per cent of the population over 65 years of age. The average inpatient stay of 41 days for these patients can be contrasted with an overall average of 7 to 10 days, and at an average hospital cost of £2,535, compared to £849 for acute care and £596 for maternity admissions. How will developing countries cope with a similar trend in resource allocation? Bicknell and Parks opine: "The complex, costly and substantially unsolved problems of support facing senior citizens in developed countries will also face the developing world."[39] One answer, they propose, citing Great Britain's explicit rationing of access to chronic renal dialysis, is to make a concentrated effort in decision-linked research, planning, and social services support to develop lower-cost alternative strategies that are effective and humane, and that inevitably will involve a rationing of scarce resources. But they imply that anything is better than the risks of a haphazard approach.[40] To quote from the WHO report, *Health of the Elderly*: "Many of the illnesses and disabilities that were formerly considered inevitable in the elderly can now be regarded as remedial. More emphasis is being given to home care as an alternative to institutional care, and improvements in housing are considered to have a central place in strategies for changing the pattern of long-term care services."[41] If the culture of the extended family unit is preserved and strengthened, then care for the elderly, the infirm, and the chronically ill is feasible, subject to the condition that the home environment must be satisfactory. Judging by past history, this is unlikely to be achieved.

APPROPRIATE HEALTH CARE SYSTEMS

Contrasting Health Care Needs

I turn now to global issues of health and to the question of the appropriateness of the Western model of health service delivery system in relation to those global needs. I will begin by considering the principal differences between the more developed and less developed countries in terms of their health requirements.

The demographic trends in the latter are high birth-rates, low death-rates, and falling fertility rates, leading to a rapid growth of the population (2.1 per cent), with nearly 40 per cent of the population under 15 years of age and a very low proportion of elderly people (4 per cent). The less developed countries have a fertility pattern that is excessive and tragically wasteful in terms of both maternal and child health, in direct contrast to the industrialized countries, which have a low fertility pattern with very little maternal and child wastage. The industrialized countries, on the other hand, have an almost static or slow-growing population (0.5 per cent) with less than a quarter (22 per cent) under 15 years and a very rapidly growing aged population of 12 per cent (about half the proportion of young and three times the proportion of the elderly compared with the less developed countries).

The wealth of the less developed countries is still largely agrarian, based mainly in non-monetary subsistence farming. Cash crops are frequently surplus to world purchasing practices. While the economies of some developing countries are growing faster than those of industrialized countries, the per capita growth rate offsets this differential. The huge difference in per capita incomes (U.S.$16,000 as compared to U.S.$700) and a differential income growth rate ensures that the adage "the rich grow richer and the poor grow poorer" remains true. The less developed countries have severely limited financial resources and low per capita incomes. While degradation of the soil continues apace in developing countries, the industrialized countries accelerate their pollution of the environment. The richer countries have extensive primary, secondary, and postsecondary educational facilities and manpower, whereas the poorer countries are still struggling to achieve 100 per cent primary school education with a paucity of trained teachers. Developing countries have 25 per cent of the world total of enrolled secondary school students; the developed countries have 75 per cent. Postsecondary school education is obtained by around 250 per 10,000 persons in the underdeveloped countries, and by only 40 per 10,000 in the less developed countries. (These figures are inexact, but they represent truly the magnitude of the difference.)

While the epidemiological pattern of disease differs substantially from that in the developed countries, the developing countries appear to be following the same pattern from high endemicity to low endemicity interspersed with epidemics, and from acute diseases with high morbidity and mortality in the young to a chronic disease pattern affecting middle-aged and elderly people. The pattern changes from the predominance of the communicable, infectious, parasitic, and vector-borne diseases with a heavy incidence on the alimentary and respiratory systems to one of stress and degenerative afflictions affecting the cardiovascular and nervous systems, e.g. heart disease, strokes, Alzheimer's disease, diabetes, cancer, and so on. For instance, Jamaica, which but a few years ago exhibited a disease pattern typical of underdevelopment, now reports cardiovascular disease and diabetes among the ten leading causes of death.

The nutritional picture changes from one of undernutrition and malnutrition, both specific and general, acute and chronic, to one of overnutrition with associated obesity and diabetes. Mental diseases, alcoholism, drug addiction, and trauma affect both under- and over-developed countries. Culturally, the underdeveloped countries remain essentially agrarian societies in which the extended family unit remains strong in rural areas but breaks down in urban areas. There is a dramatic and traumatic change from communal responsibility and support to individual survival. The restraints of traditional lifestyles are overturned by the freedoms and the most undesirable patterns of urban living. The more developed countries exhibit a strong trend toward the breakdown of the "nuclear family" in favour of the "one-parent family." Thus the support of the extended family – the mini-welfare state – is lost to the concept of the impersonal national welfare state. It is against these issues that the Western model of health care has to be considered. While developing countries are essentially rural, only one-third of their population being urbanized (as opposed to three-quarters in the industiralized countries), the urban population is growing much more rapidly (4–5 times) in the less developed than in the industrialized countries, thus creating even more developmental health and social stresses. The health tools that we deploy are not in themselves sufficient but need the support of many disciplines – economic, social, agricultural and educational – to affect the transition from ill-health to health.

A Health Service Concept

At the beginning of the twentieth century, shortly after the Boer War in South Africa, a report of the Royal Commission of the U.K. on the Poor Law recommended the founding of a unified state medical service organised in local areas. The basic objectives of a public health service, according to the report, would be the provision of preventive services in order to (1) reduce illness and disability to a minimum; (2) treat illness in its earliest stages through readily available, domiciliary services; and (3) provide hospital services to support domiciliary services with specialist care that cannot be given at home.[42]

These concepts are as valid today as when they were first articulated. Writing in 1955 on the introduction of the health centre concept to Kenya, I remarked on the need "for a more extensive and more intimate approach to the African in sickness and in health," for auxiliary staff "to get into the homes of the people," for coordinating sanitary and therapeutic services, for the use of the Health Centre as a base only – in short, for taking services to the people. I recommended that such centres should be rural, to redress the imbalance between urban and rural health services; that "existing services should be integrated with the health centre concept," not vice versa; and that health professionals should "concentrate on health education all the time."[43] I still consider this to be the way forward. It is the penetration of villages and homes to take care of the *daily* health needs of families, the interaction between consumer and purveyor in the home and working environment, that provides a sound basis for change and progress.

The process of change also depends upon a recognition of the interactions between and the priorities of the various components of health care provision. It is perhaps fair to say that this process, in medical and health care, has shifted too far away from the concept of care within the family environment to care within the hospital. In fact, the hospital has become the predominant institution in the health care system of the industrialized world since the turn of the century, a trend that no doubt has been accelerated by the growth of medical science and technology and by the rise of specialities. The technologies are mere extensions of our five senses; clinical medicine can function quite logically and scientifically in many circumstances with a minimum of advanced technology. We need, perhaps, to emphasize noninstitutional care – that is, daily health care requirements within the family and community environment. And we need to consider what specific technologies are necessary for that purpose, and to marshal such technologies within the communities so that they are primarily, even exclusively, available to all primary health care personnel in that area. Currently, in many countries those technologies are concentrated mainly within hospitals, whose personnel have priority in their use. Reform and de-institutionalization of medical care does not mean the destruction of high technology hospital care, but simply its reservation for those most in need, together with the strengthening of community care through the provision of separate diagnostic technological facilities.

It was Abraham Flexner who cautioned us in 1925 that "Science resides in the intellect and not the instrument."[44] Nevertheless, the disciplined use of the mind can be materially aided by modern diagnostic technology, whereas it tends to wither without it and to confine itself to the practice of empirical medicine. So-called first-class medicine can be practised both within and without the hospital, given adequate and appropriate support.

Table 6.7
Steps in Community Diagnosis for Curriculum Development

Social survey of KAPE*	Indicates appropriate levels of care
Epidemiological data	Determine the relevant health and sickness content
Facilities analysis	Indicates the relevant logistics necessary
Job analysis	Leads to realistic job specifications
Economic resources available	Determines priorities

The whole results in a practical and feasible curriculum.

*KAPE: knowledge, attitudes, practices and expectations of the community in relation to health services

Epidemiological and Social Analysis as a Basis for Reform

Because epidemiology is an exact science we should be able to determine precisely what our requirements are in personnel, educational preparation, facilities, and workloads in order to achieve improved noninstitutional community care. The standard of that care will, of course, vary according to the available human, economic, and social resources. I have written frequently on the potential of auxiliary health workers, especially in "simplified medicine," and noted their skills.[45] Angunawela reports on the excellence of Assistant Medical Practitioners (AMPS) in the state health services of Sri Lanka. He records that in a review of 2,344 prescriptions issued by AMPS 80 per cent were considered appropriate. He comments that in most cases a definite diagnosis was impossible because of the lack of laboratory facilities, that therapy was therefore empirical, and that "medical assistants demonstrated a competence in managing most common outpatient health problems."[46] Given their willingness to serve in rural areas, their less extensive and less expensive preparation, and their more modest remuneration, they are a worthwhile investment for strengthening community health care and could become an even greater asset through the strengthening of their facilities and support structure. For this we need "epidemiological sentinel posts" to provide data to assist in the analysis and evaluation of communities. We need information on demographics, health profiles, disease patterns, utilization patterns, facilities available and desirable, personnel requirements, and skills needed. We need sociological profiles with respect to culture, knowledge, attitudes, practices, expectations, and community aspirations for improved health. Most of all, we need to know what levels of care can be afforded for each level of economic resource (Table 6.7). Model demonstration schemes which, for example, cost seven times the available per capita health expenditure are the ultimate exercises in futility. We need sentinel posts in each region if we are to promote a scientific and logical approach to the planning of health care delivery systems. We badly need also to revise our attitude towards simplified medicine. Health care may be *simplified* by analysis, but it will never be *simple*. Ensuring that health care is appropriate, effective, and safe is a complex task. We need to determine what can be performed outside the hospital, by whom, with what preparation, with what facilities, under what circumstances, at what cost, with what outcome, and at what risk.

It is obvious that our present systems do not achieve a 100 per cent outreach. Worldwide, from the highly developed and prosperous nations to those that are disadvantaged and poor, there is an outcry of dissatisfaction. The Western pattern of health care shows a large concentration of care facilities and resources within the urban conurbations, where the majority of the population lives. But in developing countries, where there is still a concentration of facilities in the urban areas, the vast majority of the population is rural. Wherever one travels in the world the urban areas proper have relatively high quality and high density care. One is seldom out of reach of immediate care. In the rural areas and peri-urban septic fringes, not only are the health facilities of lesser quality, but they are also desperately sparse and devoid of facilities and living amenities for health personnel. Transport is deficient and uncomfortable for the sick, ambulances are notable for their absence, and the time taken to reach major health care centres is excessive and sometimes ill-afforded. The cost often outweighs the value of the services eventually received. This alone should dictate a much greater concentration upon primary health care, including health transport facilities in developing countries. We need to study the appropriateness of both static and mobile primary health care systems, by road, rail, bicycle, donkey, and shanks's pony. We should design our services in rural areas to make appropriate and modern health care readily accessible by the many. We need a system that actively reaches out to meet the sick, not one that passively waits for the sick to come trudging in.

Appropriate Manpower

I do not believe that the design, distribution, staffing patterns, and high technology of Western health care are appropriate to or affordable by developing countries as a first line of intervention. Their role should be restricted to that of supporting a much more extensive outreach of primary health care. Effective health services require a much broader team approach involving many disciplines. In 1982, in a review of primary health care development involving some seventy developing countries, the WHO stated: "Globally in 25 least developed countries average public health expenditure was U.S.$2.60 per head per year; and in a further 85 U.S.$17 per head."[47] Have we ever attempted to design health services appropriate to these financial resources? The review continues to the effect that to increase annual expenditure by $10 per capita, to an average of U.S.$12.50 per head would cost U.S.$50,000 million annually until the year 2000. Even if 80 per cent of this were to be found within the developing countries themselves, the deficit would still be "three times the present level of international transfers."[48] That sum would exclude capital costs as well as the operational and maintenance costs of water, waste, and intersectoral aspects. The monetary resources that have been available have not been shared equally between rural and urban peoples. Nor have we been culturally sensitive. We have rejected or ignored the potential of indigenous medical systems. We have not cultivated coordination of care between the systems. We have been slow to encourage and support indigenous medical education systems and to help modernize their therapeutic armamentaria. Yet throughout the world these indigenous systems persist, signifying a demand from the people that we have failed to meet. Folk practitioners exist in a wide variety of categories, with numbers mostly exceeding, or at least equalling, those of their "modern" counterparts.[49] They persist despite the advantages modern medicine has to offer. But modern medicine has failed, and, failed abysmally, to achieve 100 per cent outreach, largely as a result of high costs, educational deficiencies, and cultural resistance to change.

Folk practitioners are available, many are able and willing to learn, they are resident amongst the people, they are culturally attuned to the community, they have a respected social status, they often have a heritage of a family occupation through generations, they are trusted and accepted, they accept their rewards from the community in cash or in kind, and they often work for free. An extensive and intensive learning program worldwide in the practical skills and therapeutics of modern medicine embracing local common illnesses, their causes, diagnosis, treatment, and prevention is to my mind an immediate, practical, and feasible solution that *can* be afforded. Even in the developed countries there is a strong trend towards the deployment of less intensively trained and less costly health personnel to relieve the strain on human and financial resources. With health budgets expending some two-thirds of their financial resources on personnel, this is the major prospect for achieving substantial savings for redeployment. A paper by Robinson and Larsen[50] on the potential benefits and problems associated with auxiliary community health workers is revealing of the personal and social factors that influence motivation and performance of local auxiliary health workers. Thailand reports 14,000 registered traditional doctors and a further 50,000 unregistered traditional healers, mostly residing in villages (population 56 million), as opposed to around 9,000 allopathic physicians practising predominantly in the urban areas. India has approximately 350,000 conventional physicians, 160,000 registered Ayurvedic practitioners (the Hindu system), a further 40,000 others (Unani [ancient Greek] and Siddha), and 110,000 homeopathic practitioners. All together these traditional practitioners provide a significant potential resource. They also have an approximately equal number of medical colleges as the modern physicians. There thus exists a considerable potential reservoir of traditional personnel for primary health care work, consisting of individuals who have a health orientation and traditional training, who are well-distributed demographically, who have a historical and cultural knowledge of their peoples, and who are supported through the private purse of the community. They also have appropriate moral, ethical, and social vocational attitudes towards their communities.

The approach of the Western model has been to supplant, not to supplement and complement, traditional systems and personnel. Surely there is a case for a coordinated or combined approach, not merely by utilizing traditional health practitioners in the government health system, but by actively supporting the traditional system and the private sector, e.g., by subsidizing modern drugs, clinical facilities, transport, upkeep, and, above all, educational opportunities.

The need to retard the migration to the towns demands a rural system that accords with village culture and desires; for example, home maternity care and home care for the aged is an expressed wish in many communities within developing countries. Much of the demand for medical care arises from ignorance of modern health care, therapeutics and diagnosis, and thus a lack of home self-care. What better approach than to reinforce the expertise of traditional healers with modern knowledge and to utilize their potential for health education within the village where they are respected and trusted? We could effectively transfer much of primary health care – both general and specific – from government services to the private sector by providing supplementary education and reinforcing the utilization of traditional practitioners. Strengthening and modernizing the education and training of traditional practitioners in areas such as maternity, pharmaceutics, child development, minor medical, surgical and obstetrical emergencies, immunizations, etc., could materially lessen the burden on institutional care, and materially reduce the incidence of mishaps.

SHELTER AND PUBLIC HEALTH

"Whatever the merits of the site and service approach, the countries of the North do not have more experience of it than the people in developing countries, in fact it is not known that any industrialized country has ever officially used this approach to housing and service provision." So wrote Okpala on human resettlement.[51] He advocates, as do I, that local labour and techniques should be used, and that where local materials are in insufficient supply the means of increasing their availability must be found. It appears to me that we have tried to go forward too quickly and have concentrated upon improving the materials for housing rather than ensuring that existing types of housing provide the space and living amenities for a family. For example, it is easier and structurally sounder to advocate oval mud and wattle huts as a first step to replace rondavels rather than to proceed directly to square ones with foreign building materials; or to advocate the use of multiple huts rather than multiple-room dwellings. The issue in housing should not be to change from indigenous building materials and shapes to permanent materials, but to change to more adequate designs inside and outside the dwelling, improved ventilation, improved lighting, safe and convenient sanitation, accessible and safe water, and convenient solid waste disposal. Anyone who has used primitive pit latrines and lived under corrugated iron roofing in the tropics will agree as to their dangers to health and comfort. Indigenous building materials represent appropriate technology, which should be modernized rather than replaced.

It was McMillan in *Africa Emergent* (1949) who stated that "Africa needs great general practitioners who are humanists rather than experts."[52] This is still true today in all rural areas of the world. Medical practitioners and health officials in rural areas need to be generalists rather than specialists, able to work with a minimum of facilities and support and as a member of an ecological team; and consultants need to be super-generalists, not super-specialists!

CONCLUSION

Population programs are aimed at achieving a more equable balance between population growth rates and diminishing resources, and at accelerating the growth rates of economies so that improvement in

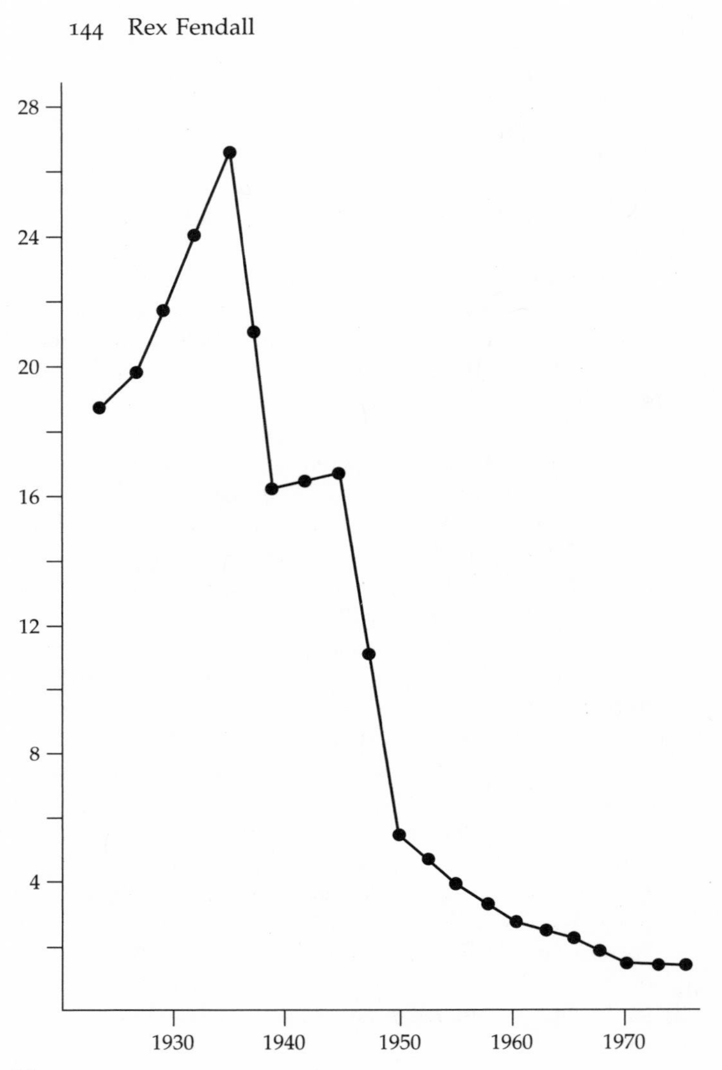

Figure 6.6
Maternal mortality per 1,000 births, Sri Lanka, 1925–75. *Data source*: K.N. Seneviratne, "Some Aspects of the Economics of Health Care in Sri Lanka," unpublished.

living standards and the quality of life can result in an acceptable level of health for all. Present trends of population growth are resulting in a shrinkage of biological resources, depressing per capita economic growth rates and effectively contributing to a worsening of global health. The problem of ill-health cannot be dealt with successfully in isolation. Nor can economic growth solve the problem of ill-health unless there is a parallel social development. I include two graphs by

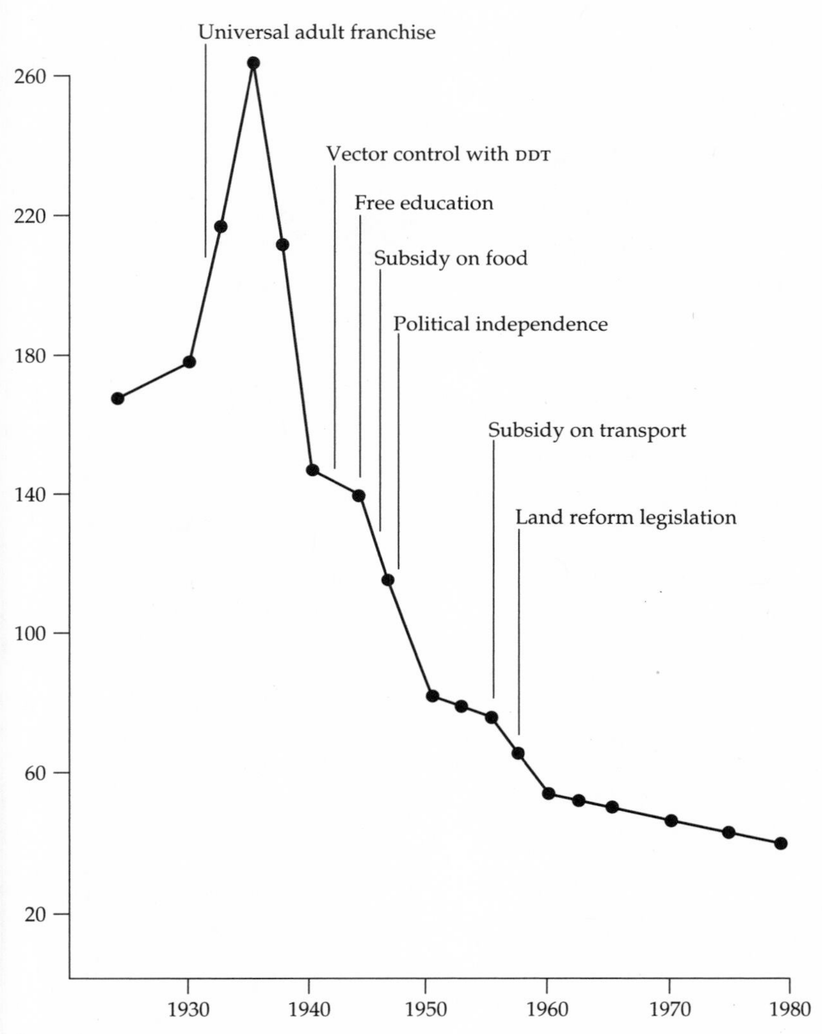

Figure 6.7
Infant mortality, Sri Lanka, 1920–79. *Data source*: K.N. Seneviratne, "Some Aspects of the Economics of Health Care in Sri Lanka," unpublished.

my late colleague, Professor K.N. Seneviratne of Colombo University, Sri Lanka, relating the fall in infant and maternal mortality to some significant steps in social reform (Figures 6.6 and 6.7). I would aver also that improved health makes a serious contribution to socio-economic development: good health care has its "profit outcome." To a public health physician, hospitals are, sadly, a visible reproach against the failure of public health provision to prevent illness.

In the long run the public's ill-health is a social responsibility, even if in the short run it is the physician's responsibility to relieve the suffering of individuals. Public health is certainly not exclusively a medical responsibility. Public health will respond only to programs of social reform that view the world – its human inhabitants *and* its resources – as one. Part of that reform must be to ensure a balance between population and resources. And that is the responsibility primarily of politicians, but also of us all.

NOTES

1 Barlow, K. 1988. *Recognising Health*. London: The McCarrison Society, 1–11, 44–53.

2 *Health*. World Health Organization Monograph Series 7: 93.

3 United Nations 1954. *Report of International Definitions and Measurements of Standards and Levels of Living: Report of a Committee of Experts*. New York: United Nations.

4 *World Population Data Sheet 1990*. Washington, D.C.: Population Reference Bureau.

5 Walton, M. 1990. Combating poverty: experience and prospects. *Finance and Development* (International Monetary Fund and World Bank) 27(3): 2–5.

6 Rao, P., Sastry, G. 1990. Measurement of trends in health status. *World Health Forum* 11(1): 1–135, 91–2.

7 Camdessus, M. 1990. Aiming for high quality growth. *Finance and Development* (International Monetary Fund and World Bank) 27(3): 10–11.

8 United Nations Fund for Population Activities 1990. *The State of the World Population*, ed. Nafis Sadik. Report. New York: UNFPA, 1–34, 4 (Box 1).

9 UNFPA 1990, 4 (Box 1).

10 Ibid., 7.

11 Ibid.

12 Gunatilleke, G., Wanigasekera, E., Namasivayam, P., and Asirwatham, S.R. 1984. *Intersectoral Action for Health: Sri Lanka Study*, ed. V. Gunawardena. Colombo, Sri Lanka: Marga Institute, 73.

13 Yi, Z. 1990. *The Fertility Revolution in China*. Report, International Development Research Centre, Ottawa 18(4): 16.

14 UNFPA 1990, 9.

15 Ibid.

16 Macoloo, G.C. 1988. Housing the urban poor. *Third World Planning Review* 10(2): 159–174.

17 Lee-Smith, D., Menon, P.A. 1988. Institution development for delivery of low income housing. *Third World Planning Review* 10(3): 217–238.
18 Wolman, A. Hollis, M.D. 1971. *Boletin Oficina Sanitaria Pan Americana* 71(3).
19 United Nations Conference on Human Settlements 1976. *Report of Habitat.*
20 Rotival, A.H. 1990. Beyond the decade. In *Developing World Water*, ed. J. Pickford. London: Grovenor Press, 1–468, 448.
21 WHO 1982. *Review of Primary Health Care Development*. SHS/82.3, 1–359. Geneva: WHO, 163.
22 UNFPA 1990, 4 (Box 1).
23 Gunatilleke et al 1984; see i, 11–15, 62–75.
24 Ibid., 67.
25 United Nations Conference on Human Settlements 1976; Shattock, F.M., Fendall, N.R.E. 1981. Restraints and constraints to development: an overview. *Rural Life* 26(3): 3–11, 9.
26 Bhatia, J.C. 1990. Light on maternal mortality in India. *World Health Forum* 11(2): 188, 191.
27 WHO 1982, 163.
28 Ibid.
29 UNFPA, 4 (Box 1).
30 See UNFPA 1990, "Social Indicators" (tables).
31 Fendall, N.R.E. 1980. The relevance of medical education to health needs. *Ceylon Medical Journal* (Sept.–Dec.)
32 Fendall, N.R.E., Marwah, S.M. 1980. Inter-university collaboration: community health. *Overseas Universities* 27: 36–41, 39.
33 World Bank 1990. *Poverty.* World Development Report. New York: World Bank/Oxford University Press, 242, 172.
34 WHO 1982. *Review of Primary Health Care Development*. SHS/82.3, 71, Table 11; World Bank 1990, 232–3, Table 28.
35 See UNFPA 1990, "Social Indicators" (tables).
36 WHO 1989. *Health of the Elderly.* Technical Report Series no. 779, 1–98, 14.
37 Bicknell, W.J., Parks, C.L. 1989. As children survive. *Social Science and Medicine*. 28(1): 59–67, 59.
38 Bicknell and Parks 1989, 61.
39 Ibid.
40 Ibid. 63–64.
41 WHO 1989, 9–10.
42 Royal Commission on the Poor Law.
43 Fendall N.R.E. 1955. Rural health centres in North Nyanza, Kenya. *Journal of Tropical Medicine and Hygiene* 58(6–7): 123–32, 149–57.
44 Flexner, A. 1925. *Medical Education*. London: Macmillan, 5–7.

45 Fendall, N.R.E. 1972. *Auxiliairies in Health Care*. Baltimore, Md.: Johns Hopkins.

46 Angunawela, I. 1990. Medical assistants as drug prescribers. *World Health Forum* 11(1): 90–1, 91.

47 WHO 1982, 53.

48 Ibid.

49 Fendall, N.R.E. 1985. Myths and misconceptions in primary health care. *Third World Planning Review* 7(4): 307–22; Fendall, N.R.E. 1981. Ayurvedic medicine and primary health care. *Tropical Doctor* 11: 81–5, 81.

50 Robinson, S., Larsen, D. 1990. The relative influence of the community and the health system on work performance. *Social Science and Medicine* 30(10): 1041–8.

51 Okpala, D.C.I. 1990. The roles and influence of external assistance in the planning, development and management of African human settlement systems. *Third World Planning Review* 12(3): 205–229, 227.

52 McMillan, W.M. 1949. *Africa Emergent: A Survey of Social, Political and Economic Trends in British Africa*. Edinburgh: Penguin, 229.

7 Environmental Toxicology

GABRIEL L. PLAA

INTRODUCTION

In its broadest definition, toxicology is the science that is concerned with the adverse effects of chemical or physical agents on biological organisms. Environmental toxicology is the branch of toxicology that deals with the adverse biological effects of chemicals found in the environment. Some might also include physical agents in this definition, but this is a relatively minor point. It is more important to emphasize that the science of toxicology is concerned with both the toxicant, which is looked upon as an "aggressor agent," and its "target," upon which it exerts an adverse effect. That target is a biological system; hence, toxicology is a biological science and draws greatly on other biological sciences. Thus, environmental toxicology is by nature a biological discipline. Interest in the presence of chemicals in an environmental setting – air, soil, or water – is not sufficient in itself to qualify as environmental toxicology, unless this observation is linked to an eventual *effect* of the chemical on a biological target.

Biological targets include humans as well as other species. In our egocentric fashion, we humans place much emphasis on ourselves as potential biological targets, but we must not lose sight of the fact that chemicals can have an important impact on other biological targets. While humans are considered a target of particular interest, other terrestrial and aquatic species are of considerable importance as potential biological targets. Environmental toxicological problems worthy of societal concern are not limited only to those that affect humans.

Ecotoxicology

There is some confusion about the discipline called "ecotoxicology." Ecotoxicology is a subdivision of environmental toxicology. Although some individuals use the two terms interchangeably, ecotoxicology has some particular characteristics that distinguish it from traditional toxicology. Ecotoxicology is concerned with the toxic effects of chemical and physical agents on living organisms, notably on populations and on communities within defined ecosystems; it includes the transfer pathways of those agents and their interactions with the environment.[1]

Ecotoxicology differs from traditional toxicology in several ways. In traditional toxicology we normally deal with the effects of a potentially toxic agent on *individual* organisms: patients, workers, domestic animals, etc. Only rarely do we consider the impact on *groups* of individuals represented as communities or populations. Furthermore, traditional toxicology certainly does not deal with interactions between multiple biological species as found in ecosystems. Ecotoxicology is concerned with the impact of toxicants on populations of living organisms or on ecosystems. Finally, the toxicological assessment procedures that were developed for studying potentially adverse effects of chemicals on individual subjects do not lend themselves well to ecotoxicological problems. Ecotoxicological problems of import require multidisciplinary scientific expertise if they are to be solved. This type of scientific interaction is difficult to achieve; as a consequence, we cannot expect major scientific strides in ecotoxicology with our current approaches.

Environmental toxicology and ecotoxicology occur in a dynamic rather than static setting. There are important interactions between the *environment*, the *target organism*, and the *chemical aggressor*. The chemical can alter the environment or the target organism; the target organism can alter the chemical or the environment; and, finally, the environment can alter the chemical or the target organism. The dynamics of the situation can have important repercussions on the overall environmental impact. In traditional toxicology, the target and the toxicant are usually well identified and information about exposure conditions is usually available. The consequences for the single individual can usually be determined. On the other hand, in environmental toxicology the targets and the toxicants are normally poorly described, and little information is available on exposure conditions. Thus, in ecotoxicological terms, consequences on populations or communities, let alone ecosystems, are largely unknown.

NATURE OF THE POLLUTANTS

Common Pollutants

A pollutant is a substance that occurs in the environment as a result of human activity, and which has a deleterious effect on living organisms. The primary sources of pollution are related to the following human activities: utilization of energy sources, increased agricultural productivity, and industrialization. Energy sources, particularly fossil fuels (Table 7.1), are certainly the largest contributor to worldwide pollution; the next greatest source is agricultural activity entailing deforestation, increased use of fertilizers, and the use of pesticides. Industrialization has brought about the synthesis of a great number of chemical entities, many of which can escape into the environment.

Pesticides are now known to be an important contributor to worldwide pollution of water and soil. Some examples are given in Table 7.2, where the agents are classified according to their "persistence." This term refers to the time required for 75 to 100 per cent disappearance of pesticide residues from the site of application. Generally, nonpersistent substances last about 1 to 12 weeks, moderately persistent agents about 1 to 18 months, and persistent chemicals about 2 to 5 years.[2] The more serious pollutants are those that are the most persistent.

Examples of important non-pesticidal organic chemicals associated with pollution are listed in Table 7.3. Many of these agents have been detected in the Great Lakes or in the Love Canal.[3] Together with the persistent pesticides, they represent the major pollutants of concern.

Chemical Characteristics of Importance

The sensitivity of the methods of chemical analysis available today is truly remarkable. We can quantify the presence of chemicals at the "parts per billion" and even "parts per trillion" level of concentration. It goes without saying that proper chemical identification at these levels requires extreme care and accuracy. We can now detect chemicals in amounts that are devoid of biological activity. Therefore, it is important that we differentiate between the mere presence of a chemical and its biological availability. A toxicant must be absorbed by a living organism and arrive at its target site (organ, cell, enzyme, etc.) before it can exert a deleterious action on the organism. The mere presence of a potentially toxic chemical in a given environmental setting is not in itself sufficient to exert a toxic effect; it must be

Table 7.1
Pollution Associated with Petroleum

Activity	*Site polluted*	*Pollutant*
Extraction	Ocean	Crude oil
Transport	Ocean	Crude oil
Refining	Atmosphere	Organics
	Waterways	Sulfur dioxide
		Mercaptans
Consumption	Atmosphere	Sulfur dioxide
		Nitrogen oxides
		Hydrocarbons

Source: Ramade 1979

Table 7.2
Classification of Pesticides in Terms of Persistence

Degree of persistence	*Pesticide*
Persistent	Chlorinated hydrocarbon insecticides
	Cationic herbicides
Moderately persistent	Triazine herbicides
	Phenyl urea herbicides
	Substituted dinitroaniline herbicides
Nonpersistent	Phenoxy acidic herbicides
	Phenylcarbamate herbicides
	Carbanilate herbicides
	Ethylenebisdithiocarbamate fungicides
	Synthetic pyrethroid insecticides
	Oreanophosphorus insecticides
	Carbamate insecticides

Source: Menzer and Nelson 1986

bioavailable as well. Thus, the establishment of the biological site of action and of the "dose" at the target site are very important for proper toxicological interpretation and predictive extrapolation.

Chemicals present in an environmental site are characterized in terms of their potential for translocation in air, water, and soil. *Bioavailability* is the key factor that must be determined in order to set priorities and to determine urgency. Chemicals that are poorly degraded by abiotic or biotic pathways exhibit *environmental persistence* and thus can accumulate. Furthermore, lipophilic substances tend to *bioaccumulate* in body fat, resulting in tissue residues. When the toxicant is incorporated into the food chain, *biomagnification* occurs as one species feeds upon others and concentrates the chemical.

Table 7.3
Nonpesticidal Organic Chemicals Associated with Pollution of Water and Soil

Aromatic halogenated hydrocarbons	Polychlorinated biphenyls Chlorophenols Dioxins
Low-molecular-weight haloalkanes	Chloroform Bromodichloromethane Dibromochloromethane Bromoform 1,2-Dichloroethane
Phthalate ester plasticizers	Di-2-ethylhexylphthalate Di-n-butylphthalate
Metals	Mercury Cadmium Lead Arsenic

Source: Menzer and Nelson 1986

Chemicals in an environmental site must be characterized in terms of their persistence, their bioaccumulation, and their biomagnification following entry into the foodchain before one can determine the nature and magnitude of the potential biological consequences. The pollutants that have the widest environmental impact are poorly degradable, are relatively mobile in air, water, and soil, exhibit bioaccumulation, and exhibit biomagnification. It needs to be emphasized once again that environmental toxicology is primarily a biological discipline. Thus, the biological component has to be well understood before we can arrive at conclusions regarding the impact of pollutants on potential environmental targets.

TOXICOLOGICAL PRINCIPLES THAT APPLY TO ENVIRONMENTAL TOXICOLOGY

Dose–Effect and Dose–Response Relationships

In the sixteenth century, Paracelsus elaborated the thesis that all chemicals possess toxic properties and that the dose determines whether or not toxicity will occur. Thus, the concept of a dose-dependent relationship in toxicology is not a modern idea. The dose–effect and the dose–response relationship, as applied by toxicologists today, mean that as one increases the dose, the appearance of adverse effects, their severity, and the number of subjects exhibiting toxic effects will increase; more importantly, it means that exposure conditions can be found where toxicity will not occur.

Toxicity may be acute or chronic.[4] Acute toxicity is the result of a single exposure or a few repetitive exposures to a chemical. Chronic toxicity is the result of multiple, repetitive exposures. The signs and symptoms exhibited in chronic poisoning are frequently different from those seen in acute poisoning. Consequently, before one can say whether or not the presence of a chemical is potentially adverse, we must know (a) if it is present in amounts that are sufficiently large to exert a toxic response and (b) the frequency of the exposure.

Toxicologists also make a clear distinction between the terms "toxicity" and "hazard."[5] Toxicity is a qualitative term that applies to all agents: *all* chemicals possess toxic properties. For some, this effect is observed when very small quantities are absorbed; for others, large amounts must be absorbed to exert an effect. Hazard, on the other hand, is the likelihood that the toxicity will occur given the manner in which the chemical is being used. To assess hazard, one needs to know the toxic properties of the chemical and the amounts available to the organism. The conditions of exposure will determine the amount of material absorbed by the organism. Thus, biological organisms, including humans, can be exposed to potentially toxic substances without hazard, if the dose absorbed is sufficiently small and the exposure is nonrepetitive. One can be exposed to subtoxic doses repetitively without exhibiting cumulative toxicity.

The Threshold Concept

While the concept of "thresholds" is of major importance in modern toxicology, its origin may be traced to the observations and writings of Paracelsus (all substances can be poisons, but the dose determines what is *not* a poison). The dose–effect relationship is crucial to toxicology. It is the underlying basis for establishing nonhazardous exposure levels in situations where humans and other mammals are likely to come into contact with potentially toxic chemical agents. A toxicant must be present at its cellular site of action in sufficient concentration to exert its deleterious effect. When the concentration is too small, we say that the threshold has not been reached. An interesting stochastic analysis indicating that "thresholds" (critical concentrations) appear to be essential for all mammalian processes has been published.[6]

There are a number of mechanisms that can lead to the demonstration of the existence of thresholds. The classical receptor agonist–antagonist relationships observed in pharmacology represent one example. Pharmacokinetic thresholds also exist, where absorption, distribution, biotransformation, and excretion can determine the effective "dose" of the potential toxicant at its biological target site. Even the tissue response itself can act as a type of threshold. The latter is evident in the development of cirrhosis in humans following long-term exposure to ethanol. In this situation, hepatic regeneration after repetitive liver injury eventually results in the production of fibrous tissue rather than normal hepatocytes. From these and other examples in the literature, it is evident that a nonhazardous situation can be achieved in a variety of ways with potentially toxic agents.

From a regulatory standpoint, the threshold concept has had important practical consequences. Safe exposure conditions for humans coming in contact with various chemical agents could be established. With pharmaceutical products this has led to the development of beneficial therapeutic interventions without accompanying toxic effects. The use of ventilation systems to avoid buildup of harmful concentrations of solvents in ambient air in the workplace is another example. The establishment of permissible chemical residues in food products is common practice. The application of this approach to drinking water is also well established.

The applicability of the threshold concept to chemically induced cancers, however, has been questioned. Some scientists maintain that the nature of the carcinogenic process is such that one cannot assume the presence of a threshold. With the so-called genotoxic carcinogens, i.e. agents that act on DNA, this has been particularly the case. Most toxicologists would agree that even with this type of carcinogen one should be able to find a dosage level that would not result in neoplasia; there would be considerable division of opinion, however, on its predictability. On the other hand, with regard to non-genotoxic carcinogens, the consensus favours the position that dosages which will not result in neoplasia can be estimated reasonably well. As mechanisms of action become better understood, scientists are more likely to apply the concept of thresholds to the regulatory process. Recent events suggest that a threshold for dioxin, based on the occupation of Ah receptors, might be considered as a possible means of developing a receptor-based model for dioxin risk assessment.[7]

With respect to another environmental pollutant, lead, there is a shift in thinking regarding the applicability of a practical threshold.[8] Epidemiological data for behavioural deficits in young children are said to be consistent with the hypothesis that deficits observed with increasing lead absorption may be without a threshold. The impact of such observations on the regulatory aspects of environmental lead exposure will be interesting to follow in the next several years.

TOXIC MANIFESTATIONS

Biological Systems Affected

Toxic manifestations depend on the severity of the exposure and on the frequency of the exposure. The general population have serious concerns about their own safety in the face of environmental pollution. People are particularly concerned about low-level, long-term exposure to potentially toxic chemicals already identified in various environmental sites. The organ systems that are of particular interest to toxicologists include the central and peripheral nervous systems, the reproductive system, the haematopoietic system, the liver, and the kidneys. As far as the Great Lakes region is concerned, there are no definite indications that serious health problems are associated with pollution, but adequate epidemiological studies are too few to support definitive conclusions. Examination of wildlife in the region, however, does indicate that several species have been affected and that occurrences of toxic reactions are attributable to the presence of chemical pollutants in this region. Toxic effects that have been documented in wildlife predators in the Great Lakes basin include: mortality, presence of malformations, eggshell thinning, effects on reproduction, and population decreases.[9] Unfortunately, information regarding the specific agents involved and quantitative knowledge about the amounts encountered is too sparse to allow meaningful conclusions to be drawn. Nevertheless, it is evident that the potential for toxic manifestations does exist.

There are three major societal fears about health and the environment. One concerns reproduction, as well as the healthy development of infants; another concerns cancer. The third is related to the fact that pollutants in the environment do not normally exist as separate chemical entities but are present as mixtures.

The Environment and Cancer

It is appropriate to discuss the significance of "environmental factors" for avoidable cancer risks. In the mid-1960s, epidemiological studies suggested that about 75 per cent or more of human cancers were potentially preventable, since these cancers appeared to be caused by "extrinsic factors," that is, by factors that were external to the biological makeup, or "intrinsic factors," of the individual. On occasion, the phrase "environmental factors" has been substituted for the phrase "extrinsic factors." Furthermore, "environmental factors" has been misinterpreted by some individuals to mean simply "man-made chemicals." As a result, we have been led to the interpretation that man-made chemical agents in the environment might be responsible for the majority of human cancers. This interpretation has not been borne out by investigative studies. In 1981 Doll and Peto published an extensive review of the causes of human cancer deaths in the United States.[10] Their estimates indicated that about 30 per cent of potentially preventable human cancers were attributable to the use of tobacco products, and that another 30 to 40 per cent appeared to be due to dietary habits and lifestyle. About 7 per cent could be attributed to chemicals found in the environment, including the workplace (4 per cent occupational hazard, 2 per cent pollution, 1 per cent industrial products). It is important to recognize that some man-made chemicals can cause cancer in humans; however, in the Doll and Peto review, these agents appeared to account for a minority of human cancer deaths. This is not to minimize the potentially adverse effects of chemical pollutants, but merely to put these agents in perspective with other extrinsic causes of preventable human cancers.

Interactions and Mixtures

It is now well established that the presence of two or more chemicals can result in a biological effect that is quantitatively different from what one would expect from the single administration of each agent.[11] This phenomenon is well known in the therapeutic use of drugs, in which patients may be required to take several different kinds of medication to control or combat their illness. When unpredictable results do occur, these are commonly called "drug interactions." Biological interactions may occur in the occupational setting as well, when the individual is exposed to several chemicals in the working environment. It is now known that interactions may occur if exposure to two or more chemicals is simultaneous or even sequential. Controlled laboratory toxicological studies indicate, however, that interactions do not *automatically* occur because of an exposure to two or more agents. Furthermore, when an interaction does occur, it is not always an exaggerated or enhanced (supra-additive) response; some interactions result in a less-than-additive (infra-additive) response when two or more chemicals are given simultaneously or sequentially.

Exposure to mixtures in the environment may result in altered toxic responses, the effect being quantitatively different from what one would expect from the toxic profiles of the individual agents.

Unfortunately, there is a paucity of experimental studies dealing with environmental mixtures. Nevertheless, some indicate that the altered toxic response may be a sum of the individual toxic responses (additive); or an exaggerated or enhanced response (supra-additive, synergistic); or less than the sum of the individual responses (infra-additive, antagonistic).

CONCLUSION

Environmental toxicology is a biological discipline in that the targets of chemical toxicants in the environment are living species. Prevention of chemical pollution of the environment is clearly the objective toward which we must strive. In realistic terms, however, this goal may be attained only for those substances that possess the highest potential for environmental impact. Knowledge about the chemical characteristics of potential toxicants is very important to the determination of their impact. Degradability, transfer pathways (through air, soil, and water), bioavailability, bioaccumulation, and transfer through food chains are dominant features of important pollutants.

The basic principles and concepts that have evolved from the more traditional branches of toxicology apply to environmental toxicology. Principles governing the activation/deactivation of chemicals and the interaction of toxicants with the biological targets are just as applicable to environmental toxicology as they are to other branches of toxicology. A thorough understanding of appropriate dose–effect relationships is fundamental to the description of a particular environmental situation. From this emanates the concept of thresholds to establish nonhazardous environmental conditions.

Toxicological studies have been used for over a century to safeguard humans from the potential harmful effects of chemicals used medicinally or in the occupational setting. The challenge now is to protect other biological species from chemical pollutants in the environment. Traditional toxicology fares reasonably well with individual targets, such as the patient taking medication or the worker exposed to chemicals in the industry. When the exposure setting becomes the environment and the biological target consists of multiple organisms, traditional toxicological approaches are less appropriate. These types of problems call upon the expertise and knowledge base of a number of scientific disciplines. In 1977, Truhaut stated: "The multiplicity of problems to be studied requires the coordination of research on a worldwide scale in a multidisciplinary context."[12] Fifteen years later his words still ring true.

NOTES

1 Truhaut, R. 1977. Ecotoxicology: objectives, principles and perspectives. *Ecotoxicology and Environmental Safety* 1: 151–73; Butler, G.C. 1978. *Principles of Ecotoxicology.* Chichester: John Wiley & Sons; Ramade, F. 1979. *Ecotoxicologie,* 2ème édition. Paris: Masson; Smeets, J. 1979. New challenges to ecotoxicology. *Ecotoxicology and Environmental Safety* 3: 116–21; Moriarty, F. 1983. *Ecotoxicology.* London: Academic Press.

2 Menzer, R.E., Nelson, J.O. 1986. Water and soil pollutants. In *Casarett and Doull's Toxicology,* 3rd edn, ed. C.D. Klaassen, M.O. Amdur, and J. Doull. New York: Macmillan, 825–53.

3 Government of Canada 1991. *Toxic Chemicals in the Great Lakes and Associated Effects – Synopsis.* Ottawa: Ministry of Supply and Services Canada.

4 Plaa, G.L. 1989. Introduction to toxicology: occupational and environmental toxicology. In *Basic and Clinical Pharmacology,* 4th edn, ed. B.G. Katzung. Norwalk, Conn.: Appleton & Lange, 733–41.

5 Plaa 1989.

6 Dinman, B. 1972. "Non-concept" of "no-threshold" chemicals in the environment. *Science* 175: 495–7.

7 Roberts, L. 1991. EPA moves to reassess the risk of dioxin. *Science* 252: 911.

8 Silbergeld, E.K. 1990. Toward the twenty-first century: lessons from lead and lessons yet to learn. *Environmental Health Perspectives* 86: 191–6.

9 Government of Canada 1991.

10 Doll, R., Peto, R. 1981. The causes of cancer: quantitative estimates of avoidable risks of cancer in the United States today. *Journal of the National Cancer Institute* 66: 1191–1308.

11 Grisham, J.W. 1986. *Health Aspects of the Disposal of Waste Chemicals.* New York: Pergamon Press; National Research Council 1988. *Complex Mixtures.* Washington, D.C.: National Academy Press; Goldstein, R.S., Hewitt, W.R., Hook, J.B. 1990. *Toxic Interactions.* San Diego: Academic Press.

12 Truhaut 1977, 170.

8 Aboriginal Peoples: The Canadian Experience

MARLENE BRANT CASTELLANO

There is a myth, recounted in various quarters, that North American Indians, aboriginal peoples native to this continent, are natural ecologists. I would like to begin this chapter by evoking a tradition which, I believe, lends credence to the myth. I will proceed from there to discuss some problems in maintaining those traditions in a modern context, and conclude with prospects which I perceive for our common future.

The Mohawk Nation, to which I belong, is one of the Six Nations of the Iroquois Confederacy, the people who lived in the Great Lakes–St Lawrence region when the earliest explorers arrived here. I begin with the greeting which was recited at the opening of Iroquois councils and as a welcome at gatherings of long-time friends as well as newcomers to Iroquois territory. The greeting speech was transmitted in the oral tradition from time immemorial, but for the benefit of those of us who have spent so much time in non-Native educational institutions that our memories have become faulty, elders in recent years have allowed the speech to be transcribed. I present here a version of the greeting received from Ernie Benedict, an elder of the Akwesasne community just east of Kingston, Ontario.

Imagine that it is early morning of a beautiful June day, that we are outdoors touching the earth which supports our feet, and that in the background are the sights and sounds and smells of the creatures who figure in the speech.

> We have come together from many directions and many distances. We have come from many houses. So we look

to the left and to the right, giving greetings to one another, for we have come together in good health and with good mind and we believe that our Creator has caused this to be so and so we give thanks and greetings to one another.

We have been instructed then to look to our Mother, the Earth, who has been supporting our feet as we travel here and there, who has been giving forth all manner of life and continues to bring those things that will enable that life to grow. We believe that our Mother Earth is faithfully following the instructions of the Creator and so we give greetings and thanks to our Mother Earth.

Upon the earth are many waters, those that are contained within bowls and lakes and those that are contained in rivers that run freely. We believe that the waters of the earth by their continued actions are necessary for all forms of life, and even though they are carrying a greater burden at this time than they have ever carried before, still they are fulfilling, to the best of their ability, the instructions of the Creator. So we give thanks and greetings to the waters of the earth.

Now upon the earth is the plant life, they who may be found between the smallest particles of the earth and who may also be the greatest creatures of the earth. For from these smallest to the greatest forest creatures, the plants are necessary to provide food, to provide medicines, to provide warmth as fuel, to provide shelter over our heads. So we believe the plant life of the earth is faithful to carrying on the duties according to the instructions of the Creator. So we give them thanks and our greetings.

Then upon the earth are the animal creatures; they also may be found between the smallest particles of earth and they also may be greater than any human in size. Now at various times we have become hungry and we have gone to the animal creatures for food. We have become cold and we have gone to the animal creatures for clothing and we believe now as we look upon our animal relatives that they also are still carrying on the instructions, each according to his own clan, his own tribal group, and the instructions of the Creator. So we look to them and we give them greetings and thanks.

Then above our heads are the bird creatures. They who fly here and there carrying the news of the changing of the seasons and singing the songs in our ears that make us joyful and contented upon the earth, they who provide food when we are hungry. So we believe that the bird creatures also are faithfully carrying on the instructions of the Creator. So we give them thanks and greetings.

Then, as we look into the sky, there is the face of our great warrior, the Sun, who is regularly making his journey across the earth, giving light, and warmth, and strength to all living creatures of the earth. He who has combined with the Earth to produce all manner of life. We believe, then, that the Sun also is carrying out his instructions according to the wishes of the Creator and so we give him thanks and our greetings.

In the path that the Sun has made, walks our Grandmother Moon. Our Grandmother Moon has the responsibility of taking charge of all the female life of the earth and indeed speaks to the human beings through their women and has also taken charge of the birth of the new life upon the earth, including the generations of human beings that are coming to us out of the ground. So we believe that our Grandmother Moon is concerned with the future generations of the people of the earth, especially the welfare of the children to the seventh generation. So we believe that our Grandmother Moon is also carrying on the instructions of the Creator and we give her thanks and greetings.

Then above the shoulders of our Grandmother Moon is a great blanket of stars. They are the oldest in creation, deserving of our great respect, and indeed the head of the proudest human when he raises his face to the stars has found that his head has become reduced to the size of that of a normal human being by his action of looking to the stars. So we believe that the stars have a special charge over the spiritual life of the people of the earth and indeed, in times past, they have watched over the travels of our ancestors. We believe that they will watch over the travels of our spirits as they enter the spirit world and go on into unknown worlds. So we believe that the star creatures also are carrying on the instructions of the Creator and we give them thanks and greetings.

Now upon the Earth is the unseen world, with the unseen people. We have felt the strength of the winds but we have not seen the wind. We know that there are great quantities of water that have come upon the earth from the rains and so we have said that the thunderers are carrying great quantities of water to replace the waters of the earth. So they also have concern with the well being of all of the life of the earth. So we give all of the unseen creatures upon the earth our thanks and our greetings.

There are among these unseen creatures certain forces that we do not understand completely and we also observe that they have great influence upon our lives. Indeed in these times, we are discovering more and more of the unseen forces upon the earth and we believe that all of these also were created to do their task and so we give them thanks and greetings.

Now as we look in all directions, we find that all of these have been brought together into the world by the Creator, who has found for himself a place where we will not see his face but where he will be listening to the words, and the prayers, of the people coming to him from out of the ground. So as we raise our voices from the earth to beyond the sky, we put our minds together as one; we look deep into our hearts and we find there the finest thoughts, and the finest words. We put all of these together as one prayer, giving thanks to the Creator above.

Now there are matters that have been brought before us and discussed for the good of our people. So on these matters we ask that the spirit of the Creator will be among us, uniting our minds and hearts.

Now those words have been spoken that come before all others. Now the meeting may begin.[1]

The Iroquois world-view reflected in the greeting speech was that of a spiritual creation interconnected in a web of responsibilities in which every creature, whether a bacterium in the earth, a songbird in a tree, or thunder in a cloud, had an essential place. Humans were neither more nor less important than any other creature in that great chain of being. I cannot imagine a child reared with that understanding suffering problems of either self-esteem or self-importance.

The kind of universe represented in the greeting speech placed ethical and behavioural obligations on human beings, and such obligations were not unique to the Iroquois Nations. To illustrate the point, I turn to another cultural group prominent in southern and central Ontario: the Anishinabek, which includes the related nations of Ojibway, Odawa, and Algonquins.

The Anishinabek, like the Iroquois, believed that they were surrounded by spirit beings who were well-disposed toward humans but who, like humans, could be offended. It was necessary then for every person to equip himself or herself for living in a world which had visible and invisible conditions. On reaching physical maturity, every young man was expected to fast and seek contact with a guardian spirit who would protect and guide him in his vocation. Powerful visions, disciplined behaviour and spirit helpers were means to the attainment of *pimadziwin*, the Anishinabe ideal of the good life – which meant a long, healthy, spiritually rich life. The rituals and ceremonies practised were intended to maintain good relations with the many spirit beings and they frequently involved giving something in return for something received. So the hunter making a shelter would ask the forgiveness of the tree which was giving up its life and would put down tobacco as a gift given to acknowledge the gift received. The obligations of reciprocity – maintaining a balance between giving and receiving – applied not only to relationships with the earth, trees, fish, animals, and winds but also to human relationships. In pipe ceremonies and ceremonies in the sweatlodge the words "all my relations" are often used as Christians use "amen" to signify the end of a prayer or act of worship. The phrase "all my relations" is a ritual acknowledgement that we are related to all the creation and that all of our relations are touched by our thoughts and actions, whether for good or ill.

I have tried to convey with words alone a sense of the ethical environment into which Iroquois and Anishinabe persons were socialized from earliest childhood. Until twenty years ago popular belief in mainstream Canada held that the cultures which fostered such beliefs and practices had all but disappeared, along with the languages in which these ancient beliefs were expressed.

It is a fact that the integrity of traditional beliefs and the transmission of knowledge in the oral tradition have suffered disruptive assault from the well-meaning interventions of missionaries, educators, and child-welfare authorities. Children separated from their parents and grandparents for ten months of the year in residential schools lost the ability to understand and appreciate their elders. Even those subjected to less depersonalizing experiences, attending day schools on reserves or lodged in boarding homes in the city, learned to devalue their origins. Many of our young people despair of finding a place for themselves in either a traditional or a competitive technological world. The most tragic stories are of those of countless children rescued from poverty and neglect by child welfare agencies only to suffer loss of identity and hope in foster and adoption homes.

The capacity of whole communities to reintegrate and reeducate their youth in traditional ways has been undermined by physical changes in the environment or wholesale displacement of settlements. In *A Poison Stronger Than Love*, Anastasia Shkylnyk documents the effects on the Grassy Narrows Ojibway of relocation from an old reserve site where traditional skills were valuable and the spirits were friendly.[2]

Despite the record of social intrusion and geographic displacement which aboriginal peoples across the country have experienced, I believe that the extent of culture loss has been greatly overstated. Five hundred years of Indian–White contact have also been five hundred years of resistance. In the final decades of the twentieth century we are seeing a widespread, deliberate attempt to reaffirm the value of aboriginal identity and to seek out traditional knowledge. What may be termed a revitalization movement has engaged many of today's young adults, with the help of elders who for decades were the silent guardians of ancient wisdom. Art Solomon, a seventy-seven-year-old Ojibway honoured for his successful advocacy of religious freedom for aboriginal prison inmates, talks of the sacred fire, *Ishkote*, at the heart of Ojibway spiritual life, which almost died out. He describes his work as that of sifting through the ashes to find the live embers which will light the fire for the next generation of Anishinabek. Jake Thomas, an Iroquois hereditary chief, ceremonialist, artist, and teacher, talks of the persecution he endured, even from his Iroquois neighbours, for holding to pagan beliefs. Nearing the age of seventy, he is a central figure in the launching of an Iroquoian Institute on the Six Nations Reserve at Brantford, Ontario, to pass on to Natives and non-Natives alike traditional teachings about the earth and the sacred responsibilities of human beings in the Creation.

The revitalization movement is calling to consciousness experiences of aboriginal culture which people like myself, schooled in rational analysis, had relegated to the sphere of the personal and subjective, assuming they had little relevance to the larger work of transforming society.

I grew up on an Indian reserve, the eighth of ten children in a family which practised a subsistence lifestyle for much of the year, growing food, fishing, and harvesting nuts and berries. Every year in late June the whole family would go out to the fields to pick wild strawberries; the group included infants carried in market baskets, and toddlers whose tiny handfuls of berries were ceremoniously added to the store of food which would be preserved for the winter. Our Anglican royalist community had long ago given up the seasonal festival of "Thanks to the Strawberry" which, in the Iroquois calendar, celebrated the first fruits of the year and the assurance of the Creator's continuing care and bounty. I discovered as an adult that these personal experiences of communal work, child-rearing, and appreciating the earth's gifts, experiences which had shaped my values and behaviour, were rooted in coherent practices stretching back centuries into Iroquois history. I became a born-again Mohawk. My heart soars when I hear the greeting speech or when I see one of my students discover the eloquence and the vision of the Great Law of Peace which is central to Iroquois history.

The rediscovery of culture and the reclaiming of identity has generated a tremendous burst of energy within aboriginal nations across Canada. The effects can be seen in a number of spheres: the assertion of political rights deriving from aboriginal status as self-governing nations; intense preoccupation in some quarters with reinstatement of ceremonial practices; widespread initiatives to incorporate aboriginal language and culture in community-controlled education; and vigorous, occasionally violent, claims to ancestral lands.

In preparing to write this chapter I turned my attention specifically to the problems and prospects of applying environmentally friendly cultural traditions to contemporary problems. I had hoped that my task would have been made easy by Anishinabek and Iroquois communities, who have been engaged for almost twenty years in struggles to assert control over traditionally held territories, and by community task forces charged with formulating strategies to secure the land for the seventh generation of children yet unborn. In reviewing recent work and position statements emanating from Ontario aboriginal communities I have come to the conclusion that we are a long way from articulating practical strategies for integrating traditional wisdom into contemporary practices of land use. Jake Thomas, my Iroquois mentor, says that the beauty of oral culture is that it is like a river, surviving over time but always changing. It is clear that aboriginal people have a tradition of understanding and respect for the complex and delicate relationships which are sustained in the environment. The record is also clear that inherited knowledge is insufficient to address problems such as the mercury pollution which contaminated the food supply in a whole river system in northwestern Ontario in the 1970s, or the drowning of thousands of caribou as a result of induced fluctuations in water levels in northern Quebec.

Until very recently the normal evolution of indigenous knowledge, i.e. adaptation to changes in the environment, has been arrested. Environmental change necessitating adaptation has been set in motion outside of the native community. It has been of a magnitude and a pace which as other chapters in this volume have shown, taxes the capacities of urban society, which thrives on innovation. However, the greatest impediment to applying traditional knowledge to contemporary problems has been the systematic exclusion of aboriginal people from strategic planning and from decision-making.

Aboriginal groups are coming to grips with the consequences of environmental degradation not as a theoretical problem but as an immediate threat to community life and health. Current initiatives by aboriginal people, as I read them, focus upon (1) participation in decision-making on land use in ancestral territories and land surrounding their residential communities, (2) jurisdiction to enact and enforce environmental regulations within their territories, and (3) the necessity of translating traditional knowledge into a contemporary language and applying it in a new context.

The exclusion of aboriginal people from decisions affecting the common environment is not peculiar to Canada. The Brundtland Report observed that:

> Many [indigenous or tribal peoples] live in areas rich in valuable natural resources that planners and "developers" want to exploit, and this exploitation disrupts the local environment so as to endanger traditional ways of life … . Growing interaction with the larger world is increasing the vulnerability of these groups, since they are often left out of the processes of economic development. Social discrimination, cultural barriers, and the exclusion of these people from national political processes makes these groups vulnerable and subject to exploitation. Many groups become dispossessed and marginalized and their traditional practices disappear.[3]

The Teme-Augama Anishnabai of northeastern Ontario have been engaged in a struggle for 115 years to assert their rights to ancestral lands. They have proposed forest stewardship principles to "integrate human uses of the forest in a manner compatible with the continuity of forest life. Forever."[4] Traditional sensibilities are evident in Chief Potts' plaintive questions: "Why should forestry companies be allowed to clear out our forests and then move on elsewhere, leaving a wasteland behind them? Where will the animals and other life forms go when the forests are gone?"[5]

Teme-Augama Anishnabai protests have been countered with indifference, injunctions, arrests, and court decisions denying any legitimacy to attempts to assert rights in the disputed territory. As has been the case with many other aboriginal groups, the Anishnabai bid to inject their perceptions of environment and responsibility have been forcibly rejected, at least until May 1991, when the new NDP government in Ontario formalized the creation of a joint stewardship council for the Temagami area.

Typically powerless to affect what happens in their region, aboriginal communities also find themselves lacking jurisdiction to protect their own reserve territories or to regulate the behaviour of their members. Anishinabek communities in Ontario have mobilized, for example, to resist dumping of hazardous waste at sites adjacent to their lands and waterways, and to secure an injunction to stop the federal government from dredging and dumping contaminated sediments in locations affecting community waterways and supplies.[6] Mike Mitchell of the Akwesasne Mohawk community cites eleven attempts by the Mohawk council to enact bylaws to secure peace and order, including a wildlife conservation by-law. In every case the federal government disallowed the by-laws, even though authority for regulation of "law and order" is granted to band councils under the Indian Act.[7]

The fact that aboriginal communities perceive the *need* to regulate their members' behaviour in matters of housing sites, local enterprise, water distribution, and sewage disposal, as well as in matters of hunting and fishing, is evidence of how their world has changed from the time when every family was responsible for its own environmental practices. In this changed universe, where father goes to work at the pulp mill instead of on the trapline, and mother earns income at the local tourist lodge rather than tanning hides to make moccasins, how will traditional ethics be transmitted to the next generation? The ineffectiveness of formal regulation was brought home forcibly to one band council which briefly authorized a waste disposal operation on its reserve. Within weeks the site had become the repository for waste contaminated with PCBs and lead, left by a haulage company under investigation by state authorities.[8]

Law, to be effective, must be written on the heart. The most imaginative environmental educators have not solved the problem of how to inculcate the deep-rooted attitudes which produce environmentally friendly behaviour in the general public. Translating respect for the environment into local regulations is one strategy being implemented, but the question of how to revive, update, and enforce the understanding of *pimadziwin*, the good harmonious life, is still problematic for many aboriginal people in this generation.

What of the prospects for the future? I see two hopeful signs. First, aboriginal personnel are overcoming their alienation from mainstream institutions and learning the language of such agencies as the International Joint Commission and the Federal Environmental Review Organization. They bring a distinctive perspective to the deliberations of those panels. Chief Robert Williams of the Walpole Island First Nation, which is located downstream from Ontario's "Chemical Valley," addressed a meeting with the Hon. Robert de Cotret, then Minister of the Environment, with these words:

> Why a "Green Plan"? Walpole Island's trees are green but our wetlands are golden. Our water is sometimes clear and our skies are often blue. Our moon is white, our night is black, and the contaminated sediments on our river and lake beds are brown. The title of the Plan badly misses the mark: it suggests a lot of "green" rhetoric which reinforces the notion that the environment is somehow separate from development. Canada does not need a plan to manage its natural environment: it needs a plan to manage equitable development in a way that preserves and improves the environment.[9]

Aboriginal people historically have been an impediment to the exploitation of natural resources, which were seen by government and industry alike as a limitless source of fuel for the Canadian economy. Aboriginal people have not abandoned their protagonist role in that respect: witness the ongoing efforts of the James Bay to force a comprehensive environmental assessment as a precondition for development of the James Bay II hydroelectric project.

We look back to our traditions not as a romantic attempt to escape present reality but rather as a way of finding touchstones to guide us into the future. Our technicians and philosopher-elders are ready and able to engage with the kind of questions raised by the Planet Earth symposium. Some people in some regions bring an intimate knowledge of ecological systems in their region, knowledge which deserves to be acknowledged as scientific in its own right, based as it is on empirical observation over long periods of time.

The second hopeful sign is that there is considerable energy within the aboriginal community, and evident support within the Canadian public, to redefine the relationship between aboriginal people and non-aboriginal Canadians. For most of the past five hundred years aboriginal people have been outsiders to the acknowledged history of this land. We have been savages: noble savages in the forest, cruel ones outside the fort. We have been backward peoples, social problems, drunks. More recently we have become victims, activists, environmentalists. All of those stereotypes contain a grain of truth, but they serve a single purpose – to cast us in a role which serves the purpose of a dominant and dominating society.

When Elijah Harper stood in the Manitoba legislature in June 1990 holding an eagle feather and saying "No" he was not simply opposing a bad piece of constitutional law. He was saying "No" to at least four centuries of subordination, depersonalization, and exclusion which have characterized our relations with mainstream Canadian society.

I believe that we have the possibility now of redirecting history. In place of the imperialist, expansionist, exploitative ethos on which the Canadian nation and Canadian character have been built, it is possible for Canadians to establish a new contract with the original people of this land and, in doing so, to plant more deeply their own earthy roots. If this happens then perhaps Aboriginal traditions of environmental relations will become a conscious and valued component of every Canadian child's heritage. Jerome Bruner, an educator lamenting the loss of compelling myths in North American culture, wrote:

> There is in the "mythologically instructed community," a corpus of images and identities and models that provides the pattern to which growth may aspire – a range of metaphoric identities … . [Such a] community provides its members with a library of scripts upon which the individual may judge the play of his multiple identities … . Myth serves not only as a pattern to which one aspires, but also as a criterion for the self-critic.[10]

Some aboriginal people know consciously, some know intuitively, and some are just discovering that the universe is made up of "all our relations." We are actively trying to translate that knowledge into practice, but to do that we need to be full partners in society. We need, as all Canadians do in this decade, to find powerful myths which will mobilize us and our children to fulfil the responsibilities which will secure Our Common Future.

Nyaweh. Thank you.

NOTES

1 The Greeting Speech, Iroquois Oral Tradition, transcribed from a presentation by Ernest Benedict, October 1987.

2 Shkylnyk, A. 1985. *A Poison Stronger than Love*. New Haven: Yale University Press.
3 The World Commission on Environment and Development 1987. *Our Common Future* (Brundtland Report). Toronto: Oxford University Press.
4 Potts, G. 1989. In *Drumbeat, Anger and Renewal in Indian Country,* ed. Boyce Richardson. Toronto: Summerhill Press, p. 208–9.
5 Potts, G. 1988. Speech delivered at the Wendaban Line, 1 June. Unpublished.
6 Union of Ontario Indians 1990. *Anishinabek Environment Policy.* Discussion paper. Toronto. Unpublished; Walpole Island First Nation 1990. *Proposal for Preparation of a Sustainable Development Strategy.* Walpole Island Heritage Centre, September.
7 Mitchell, M. 1989. In *Drumbeat, Anger and Renewal in Indian Country,* ed. Boyce Richardson. Toronto: Summerhill Press, 128.
8 Akwesasne Task Force on the Environment 1991. *Eddie Gray Recycling Facility Controversy.* 1(1): 8.
9 Walpole Island First Nation 1990.
10 Bruner, J.S. 1962. Myth and identity. In *Essays for the Left Hand,* Cambridge, Mass.: Harvard University Press, 36.

9 Science and Politics on Planet Earth

MICHAEL IGNATIEFF

In my own lifetime the way we all think about our place in the natural world has been transformed. First, we now think of ourselves as a species among other species, no longer the lords and masters of creation. We have learned that our survival as a species depends on the behaviour of other species, and upon our capacity to understand and modify their behaviour and ours. Second, we understand that we live within an ecosphere, a dynamic and complex system of processes, the complexity of which we are only beginning to appreciate. Third, we have begun to appreciate that we are a dangerous species, the one animal capable of destroying the global habitat.

This erosion of our anthropocentrism is relatively recent. Modern environmental consciousness began with Hiroshima. The atom bomb made us aware that we were the first generation of our species with the power to terminate life on the planet. A further stage in the evolution of our planetary awareness came with manned space flight, and the return to earth of those first pictures of our blue planet wreathed in cloud. Those pictures made the ecosystem visible to the smallest child. When in the 1980s we discovered the hole in the ozone layer, the general public began to grasp that there were no hiding places left: every part of our habitat was threatened if any one part of it was; a tear in the invisible envelope of the ozone would have consequences from which no one could escape.

Science has played a vital role in developing the knowledge which underlies this new environmental consciousness. The present volume bears witness to the emergence of a new scientific synthesis bringing together elements of knowledge that, until now, were considered to belong to separate fields. For the first time, we are beginning to realize that population studies, toxicology, climatology, economics, demography, and the physics and chemistry of fuel consumption belong together in what might be called the new science of Planet Earth.

In some ways, we have had to relearn things we should never have forgotten. Canadians in particular are just beginning to wake up to the fact that the aboriginal peoples of this continent have always understood that our survival as a species depends on respectful management of our ecosphere. We are only now realizing the price all Canadians have paid for forgetting, ignoring, or dismissing this vast store of aboriginal knowledge. Learning survival skills will require a vast relearning. We will have to listen to the scientists, but we will also have to learn from the tribal elders.

This point could be generalized. All local initiatives, whether they pertain to rural development, population control, or green farming, depend for their success on democratic feedback. Nothing works if it is imposed from above. Sound environmental policies emerge only where policy-makers, experts, and politicians are prepared to listen to and learn from the people closest to the ground. What might be called "vernacular" knowledge is just as important as "scientific" knowledge in the struggle to save the planet.

This having been emphasized, however, one caveat should be entered. We should not set "scientific" and "vernacular" knowledge in opposition. Guilt about our condescension towards aboriginal wisdom in the past should not lead us now to oppose it to "Western science." The world's green movements have some reason to be suspicious of Western science: ill-considered applications of science have done the environment much harm. But it remains true that if science helped to get us into this mess, it can also help to get us out of it.

The Western green movement has its historical roots in Romanticism, in the cult of nature which emerged in early nineteenth-century Europe. Romanticism was often hostile to science and reason. Romanticism in the guise of guilty reverence towards aboriginal wisdom is an intellectual luxury we can ill afford.

There is another intellectual luxury we can afford still less. This is the idea that the science of the environment is sending us inconclusive or contradictory messages. From this idea flows the most important remaining alibi for political inaction on the environment: that we do not know enough. Science does remain divided on many key environmental issues. Scientists, for example, still do not agree on the meaning of long-term climatological trends. Exactly when the full consequences of global warming will hit us remains unclear. But this is no excuse for inaction. The contributors to this volume have delivered a clear message: "We do not know enough, but we know enough to act." Science tells us we must do something now to reduce the consumption of fossil fuels and the production of greenhouse gases. It cannot tell us exactly how and when the negative effects will play themselves out. But the mandate for political action is clear.

Science has always driven the economy. It now will begin to drive politics. This is an unfamiliar role for science. It is only since the Second World War that politicians in most Western countries have begun to retain scientific advisers. In the 1990s, politicians will have to pay progressively greater attention to science and find ways to build consultation with the scientific community into the decision-making processes of governments. If they don't, the time-lag between scientific discovery and political action on the environment will become intolerable. At the moment, the relation between science and government is mediated largely by the media, or, in other words, by the fitful popular pressure which environmental groups can bring to bear on the political process. This is how it should be, and everyone can only hope that the environmental lobby grows stronger. But a democratic policy needs to do better. We need good environmental science feeding directly into the decision-making processes of politics *before* crisis turns into disaster.

Everyone realizes, however, that environmental problems pose political problems which few politicians and fewer voters feel able to confront. The soft environmental options have run their course. Recycling and energy conservation have become fashionable; most people's behaviour toward the environment has become more respectful and considerate. But in the 1990s soft environmentalism will not be enough. Harder choices about our consumption will be imposed upon us. Our very way of life will become an issue. The problem is that sensible politicians always avoid challenging their electorate's way of life.

The coming political difficulty can be illustrated by brief mention of the concept of "sustainable development." This phrase was coined by the writers of the Brundtland Report. It is a phrase which seeks to square the circle: to reconcile continued economic growth with environmental protection, extending the benefits of Western consumption to the Third World without putting the ecosystem's carrying capacity under intolerable strain.

If this volume has any simple political message to offer it is this: beware of the rhetoric of sustainable growth. The phrase *may* be an oxymoron, a contradiction in terms. It may not be so in principle, but in practice it is hard to see how rising real incomes in the Third World are achievable without substantial ecological damage. It is hard to see how growth in the West can be sustained without much deeper changes in our lifestyles than we had initially expected. The rhetoric of sustainable growth looks suspiciously like an exercise in collective self-delusion. The delusion consists in believing that soft environmentalism – more recycling, more energy conservation – will have sufficient long-term impact on global environmental indicators to make deeper cuts in our own lifestyles unnecessary.

My own prediction is that the environment will play havoc with the traditional political agenda in the 1990s. It will do so because the soft options have been exhausted and only hard choices remain. Taxes will have to rise on energy; more and more capital will have to be diverted from investment to environmental clean-up; governments will have to intervene to offset the environmental short-termism of industry and business, and they will have to push the public, through the stick of taxes and the carrot of incentives, to cut down on polluting behaviour.

Some people believe that this presents us with the looming spectre of authoritarianism. Environmental pressures will force democracies to become more and more draconian in their attempts to enforce policies of environmental survival on an unwilling populace.

My own view is less pessimistic. If anything, politicians have been underestimating the environmental awareness of voters. What I feel around me is nothing less than a sea change in the way in which ordinary people calculate the consequences of their actions over time. Once upon a time, it was possible to toss the garbage behind us and keep on walking. Now we know it washes up on our beaches. Once upon a time, it was possible to forget the consequences tomorrow of what we do today. Now we are beginning to realize that, in the case of nuclear power and the disposal of nuclear wastes, the choices we make today will weigh on our species forever. Environmentalism has dramatically extended the timeframe for which we must calculate consequences, and this is breeding a salutary caution in voters' minds about development of all kinds.

Politicians may have been underestimating the willingness of voters to make sacrifices on behalf of their children's future. My instinct is that the environment is a leadership issue. If politicians are prepared to lead, their voters will follow.

If a democratic consensus can build upon this environmental consensus, authoritarianism may not be necessary. In such a context, there is every reason to suppose that the carrot will work better than the stick. Energy pricing that encourages efficiency and tax incentives for investment in environmentally friendly production – in other words, market solutions – are bound to command more electoral support than draconian environmental edicts. Needless to say, market solutions are easier for governments to impose, and these will be the instruments of choice.

But the political message of all the parties will have to change. The environment will cease to be an add-on paragraph in party platforms. Environmental concerns will begin to dictate the whole political agenda as the environmental crisis deepens. As far as I can see, a political agenda driven by environmental concerns will be one which rediscovers the virtues of public goods and public provision.

The general public know that their private welfare can be vitiated by public squalor. Moreover, they understand the paradox that when everyone strives to increase their *private* welfare it frequently happens that everybody's *public* welfare declines. This is evident in urban transport, of course, where congestion is inexorably reducing the advantage of the private car. Every voter is aware of such diseconomies, which provide politicians with a golden opportunity to persuade the electorate to back energy-saving investment in public transport.

Growing environmental awareness, in other words, has slowly changed ordinary people's understanding of where their welfare lies. We all know now, as we did not know in the boom years of the 1960s, that we are getting poorer rather than richer as pollution reduces the quality of our life and undermines our health. We are getting poorer, not richer, if we simply transfer the cost of cleaning up after us to our children and grandchildren. These are now truisms of popular consciousness. They are the platform of awareness upon which a new environmental agenda could be built in the 1990s.

But there is an obstacle. The name of that obstacle is despair. The beginning of environmental wisdom is systemic awareness, i.e., the realization that all of the key indicators – demographic growth, fossil-fuel consumption, economic growth, cultural change – interrelate and interconnect. The problem with systemic awareness is that it tips easily into pessimism. The more you know the worse you feel. There is a substantial danger that the popular will to achieve environmental change will be eroded, rather than strengthened, by the flood-tide of alarming scientific knowledge which is bearing in upon voters from every side. Nihilism, cynicism, and despair are just as frequent responses to knowledge as action, determination, and civic courage. Scientists who play the role of Jeremiah should beware of the unintended consequences of gloomy prophecy.

The best antidote to despair is the small success story. If we look only at the global environmental picture, it will be easy to despair. If we look more closely, at small local initiatives in development, population control, and environmental planning, we can begin to tell a more hopeful tale. That is why action on a small scale is so important: the larger politics of hope can be sustained only by a multitude of small good deeds.

APPENDIX

Planet Earth Symposium: Principal's Introduction

DAVID C. SMITH
Principal and Vice-Chancellor
Queen's University

Over the past two weeks the Learned Societies have visited Queen's for their annual meetings in Canada. It has been an impressive intellectual feast for those of us who have been privileged to participate. As a welcomer for the university I have found it increasingly difficult to say something new upon each occasion, but last evening, just an hour or so before the start of the Royal Society banquet, the city unexpectedly repaved the street in front of the banquet hall. "Very impressive," one guest said to me. "Now how are you going to top that in greeting the Planet Earth Symposium tomorrow morning?" Well, I have neither a black nor a red carpet to spread before you this morning but I do wish to say that it is a very great pleasure to extend a warm welcome on behalf of Queen's to all of you joining us in this important symposium. We're pleased and honoured that it should be taking place in association with the university's sesquicentennial.

This symposium meshes well with Queen's concern to consider, as part of our 150th anniversary observance, how we can better advance learning and foster understanding of the central issues affecting our society. This symposium indeed involves a compatible meshing of the interests of all of us involved in its tripartite sponsorship.

For another of the sponsors, the Royal Society, the theme of the symposium fits into its global change program; for the Canadian Federation of Biological Societies, the theme combines well with its agenda for the next few days, particularly the Presidential Symposium of June 9th, entitled "Can Biological and Biomedical Research

Solve Major Problems Resulting From Global Change?" The tripartite sponsorship of the Planet Earth Symposium is symbolic of the breadth of considerations that must be brought to bear on a topic that is so profound, so urgent and so complex as the changing nature of our planet, Earth.

Tackling the issues requires, first, drawing together the best minds we can find nationally and internationally. It is therefore exciting to see the lineup of eminent scholars for the agenda of this conference. Scholarship is becoming ever greater in importance for practical problems. As Nathan Pusey once said, "We live in a time of such rapid change and growth of knowledge, that only the person who is in a fundamental sense a scholar, that is, a person who continues to learn and enquire, can hope to keep pace."

Tackling the issues of planet Earth requires in the second place integration of knowledge. This is difficult to achieve as specialized knowledge burgeons and intellectual investment produces a kind of proprietary bias that may be summarized in one version of an old adage: "Where one stands depends on in which discipline one sits." We must shuffle and recombine chairs through institutional devices to draw knowledge better together. I find the agenda of this symposium particularly exciting because of the way it has deliberately brought a broad spectrum of disciplinary backgrounds to bear on the issues.

Tackling the issues of planet Earth requires, thirdly, a setting-out of policy options derived from this expert and integrated knowledge. I use the words options rather than solutions advisedly. The agenda of the conference is appropriately cast in terms of questions and effects, inputs to moral judgements and political processes relevant to the preservation of planet Earth.

I shall stand in the way no longer of Digby McLaren's presentation of his fine opening paper, but I would like to quote the concluding statement from his presidential speech to the Royal Society of Canada last year: "This is the challenge of global change and the central need: a major recognition that all problems facing humankind are necessarily scientific, technical, social, political, economic, and, above all, moral and ethical." May this conference help move us along that ambitious yet essential path on which our collective life depends. Thank you for coming.

Contributors

MARLENE BRANT CASTELLANO A member of the Mohawks of the Bay of Quinte First Nation, Marlene Castellano received a BA from Queen's University in 1955 and an MSW from the University of Toronto in 1959. She has also pursued advanced studies in adult education at the Ontario Institute for Studies in Education. She has been a member of the Department of Native Studies, Trent University, since 1973, where she currently holds the rank of Professor and has served three terms as Department Chair. In 1991 she was cross-appointed to the Faculty of Education at Queen's University at Kingston.

Professor Castellano's research interests include social policy, education for development, and applications of indigenous knowledge in a contemporary context. She has acted as team leader designing college level programs to prepare health workers and early childhood educators to deliver culturally appropriate services in Native communities.

She is active as a consultant to federal and provincial government agencies, to other universities, and to Native organizations concerned with education, health and social services, and as a member of professional organizations, boards and committees.

REX FENDALL Born in Auckland, New Zealand, Dr Fendall studied medicine at University College, London, where he received a B SC in Anatomy in 1939 and qualified in medicine in 1942 (MRCS, LRCP diploma), graduating with an MBBS degree in 1943. After internship at University College Hospital and in various hospital residencies in the U.K. he entered Her Majesty's Overseas Medical Service in 1944 and was posted to the Yaba School of Medicine, Nigeria. He served the remainder of the war in the British military administration in the Far East and immediately after the war resumed his civilian career in Malaya and Singapore. In 1948 he transferred to Kenya, completing his posting in 1964 as Director of Medical Services of Kenya. During this time he pursued his academic interests in tropical medicine. In 1952 he qualified as an MD at London University College and Hospital and received a Diploma in Public Health (DPH) from the London School of Hygiene and Tropical Medicine.

Dr Fendall was appointed to the staff of the Biomedical Science Division, Rockefeller Foundation, New York, in 1964. In 1967 he was appointed Regional Director of the Technical Assistance Division of the Population Council, Inc. of New York. From 1966 to 1979 he was a visiting lecturer at Harvard University.

In 1971 he returned to England as Professor and Head, Department of Tropical Medicine, University of Liverpool. His brief was to reorient the department to modern concepts of community care, with particular regard to health care issues in developing countries. Thus it became the Department of International Health two years later. Under Dr Fendall's leadership three international interdisciplinary courses were developed: a Certificate in Tropical Community Medicine and Hygiene, a Master's program in International Community Health, and a teacher training course for Primary Health Care. These courses mirrored Dr Fendall's interests in organization and planning of health services, development of auxiliary staffing, primary health care, and epidemiology, all in relation to disadvantaged areas.

Professor Fendall retired in 1981 and was appointed Emeritus Professor in recognition of his service to International Medicine and to the University of Liverpool. However, he has remained active as a teacher and international consultant. In 1982 he was appointed Visiting Professor of Public Health, Boston University, and in 1983 became Adjunct Professor of Community Health Science, University of Calgary, positions he still holds.

Dr Fendall has been the recipient of numerous awards, including Fellowship of the Faculty of Community Medicine in 1972 and Fellow of the Faculty of Public Health Medicine, U.K. in 1989. He was a member of the Panel of Experts, World Health Organization, from 1957 to 1982 and a consultant to the World Bank, the United Nations Fund for Population Activities, the International Development Research Centre in Ottawa, U.K. Overseas Development Aid, the British Council, and numerous other international organizations.

He is author of *Auxiliaries in Health Care* (1972), which has been translated into French and Spanish, co-author of *Use of Paramedicals for Providing Care in the Commonwealth* (1979), and *Restraints and Constraints to Development* (1983). He has published over 150 articles on primary health care and epidemiology.

M. BROCK FENTON Born at Mackenzie, Guyana, Dr Fenton studied at Queen's University at Kingston, where he received a B SC in Biology in 1965. He pursued Graduate Studies at the University of Toronto, receiving an M SC and PH D in Zoology in 1967 and 1969 respectively. He was appointed Assistant Professor of Biology at Carleton University in 1969, receiving promotions to Associate Professor in 1974 and Professor in 1981. From 1986 to 1994 he served as Chair, Department of Biology, York University. In January 1995, he became Associate Vice-President (Research) at York University.

Dr Fenton has held visiting professorships at Rockefeller University, the University of Texas at Austin and Cornell University. The recipient of teaching and research awards, Dr Fenton has supervised numerous student theses. He has served on a variety of Natural Sciences and Engineering Research Council of Canada committees and study sections and was Vice-President of the Biological Council of Canada from 1986 until its merger with the Canadian Federation of Biological Societies (CFBS) in 1990. He is currently the President of the Canadian Council of University Biology Chairs, and Chair of the CFBS subcommittee on the public awareness of science.

Dr Fenton's research interests on animal behaviour, zoology, and evolution use the diversity of bats as an animal model, a system he first studied for his graduate theses. He has published over one hundred research articles and reviews and three books on bats and has appeared on CBC TV to discuss bats.

MICHAEL IGNATIEFF Born at Toronto, Dr Ignatieff studied at the University of Toronto and received a BA from Trinity College in 1969. During this period he was a reporter for the *Globe and Mail* for one year and a member of the campaign staff of the Liberal Party of Canada for the 1968 federal election. A Teaching Fellow in Social Studies at Harvard University from 1971 to 1974, he received a PH D in 1975.

Dr Ignatieff was appointed Assistant Professor of History at the University of British Columbia in 1976 and, in 1978, Senior Research Fellow at King's College, Cambridge, where he received an MA. Since 1984 he has pursued a career as a writer and broadcaster in London, England. He has been host of the BBC TV program "Thinking Aloud" since 1986 and co-host of "The Late Show" since 1989.

A visiting professor at the Ecole des Hautes Etudes, Paris, in 1985, he received the Governor General's Award in 1988.

His publications, which reflect his training as a historian and his interest in politics, include *A Just Measure of Pain: The Penitentiary in the Industrial Revolution*, *The Needs of Strangers*, and *The Russian Album*. He is the co-author of *Wealth and Virtue: The Shaping of Classical Political Economy in the Scottish Enlightenment* and of a screenplay, *1919*.

In 1990 Dr Ignatieff became an editorial columnist for the *Observer* (London).

MICHAEL B. McELROY Born at Shercock, County Cavan, Ireland, Dr McElroy received a BA in 1960 and a PH D in 1962 in applied mathematics from Queen's University, Belfast. After spending a post-doctoral year in the Theoretical Chemistry Institute at the University of Wisconsin he was appointed Assistant Physicist at the Kitt Peak National Observatory, Tucson, Arizona, in the Planetary Sciences Division, with promotions to Associate and Physicist in 1965 and 1967 respectively. In 1970 he was named Abbot Lawrence Rotch Professor of Atmospheric Sciences, Division of Applied Sciences, Harvard University, and in 1975 was appointed Director, Center for Earth and Planetary Physics, a position he held until 1978. In 1986 he was appointed Chairman, Department of Earth and Planetary Sciences, his current position. His research interests on the origin and evolution of the planets resulted in service on a number of NASA committees and participation in the Viking mission.

He has also served on the Space Program Advisory Panel, Office of Technology Assessment, U.S. Congress, and on numerous committees and panels of the National Academy of Sciences, including the Committee on Global Change. The recipient of prestigious awards such as the Macelwane Award of the American Geophysical Union, the American Association for the Advancement of Science "Newcomb Cleveland Prize" with the other participants in the Viking mission, and the NASA Public Services Medal, Dr McElroy received the George Ledlie Prize in 1989, awarded to the person at Harvard University who "since the last awarding of said prize, has by research, discovery, or otherwise made the most valuable contribution to Science, or in any way for the benefit of mankind."

Dr McElroy has published over 170 scientific papers. More recently his research has included an emphasis on effects of human activity on the global environment of the earth.

DIGBY J. McLAREN Born at Carrickfergus, Northern Ireland, Dr McLaren studied before and after the Second World War at Cambridge

University, where he received BA and MA degrees. He obtained a PH D in geology from the University of Michigan in 1951.

He spent the war years 1940–46 in the Royal Artillery of the British Army, serving overseas for three and a half years in thirteen countries. In 1948 Dr McLaren joined the Geological Survey of Canada and worked in the Rocky Mountains of Alberta and British Columbia, the District of Mackenzie, the Yukon, and in the Arctic Islands, specializing in Devonian stratigraphy and palaeontology. He took part in 1955 in "Operation Franklin," a major reconnaissance mapping project in the Queen Elizabeth Islands led by Y.O. Fortier, and contributed to the final report. He became Head of the Palaeontology Section of the Geological Survey in 1959 and moved to Calgary in 1967 to become the first Director of the Survey's Institute of Sedimentary and Petroleum Geology. He returned to Ottawa in 1973 to become Director of the Geological Survey, and, in 1980, Assistant Deputy Minister, Science and Technology, in the Department of Energy, Mines and Resources.

In international science Dr McLaren was active in the International Union of Geological Sciences between 1968 and 1981. During this period of rapid expansion of the Union's activities he headed groups of specialists from many countries examining the subdivision and correlation of sedimentary rocks in all parts of the world. From 1971 to 1981 he served on the Board of the International Geological Correlation Program which was supported by the Union and by UNESCO, and was chairman of the Board from 1977 to 1981.

In 1981 he joined the University of Ottawa as a Visiting Professor in the Department of Geology, a position he still holds. He retained his responsibilities with the Department of Energy, Mines and Resources as Senior Science Advisor until 1984. He proposed, convened and coordinated two Dahlem Conferences in Berlin in 1986 on Resources and World Development, and edited the final volume of conference proceedings, published in 1987.

He has published over one hundred scientific reports and maps in various international journals and those of the Geological Survey of Canada in the fields of palaeontology, biostratigraphy, and regional geology.

He was the recipient of an honorary degree from the University of Ottawa in 1980 and is a Foreign Associate of the U.S. National Academy of Sciences and a Fellow of the Royal Society of London. As President of the Royal Society of Canada from 1987–90 he was responsible for the successful implementation of the recently expanded role of the Society and has been closely associated with the Canadian Global Change Program initiative of the Society. Currently, he continues to pursue his research interests and to publish on problems of correlation and major extinction events in the past, as well as on resources and world development.

GABRIEL L. PLAA Born at San Francisco, Dr Plaa received a B SC in Criminalistics in 1952, an M SC in Comparative Pharmacology in 1956, and a PH D in Comparative Pharmacology and Toxicology in 1958 from the University of California. He was appointed Instructor in the Department of Pharmacology at Tulane University in 1958 and was made Assistant Professor in 1960. In 1962 he accepted the position of Assistant Professor in the Department of Pharmacology at the State University of Iowa, where he was promoted to Associate Professor in 1963. In 1968 he was appointed Professor and Chairman of the Department of Pharmacology at the Université de Montréal. From 1979 to 1989 he served as Vice-Dean for Research and Graduate Studies at the Université de Montréal. His research interests include forensic and clinical toxicology, liver injury, and the effects of drugs on liver function.

Since 1965, he has served on various advisory panels and committees on toxicology, food additives, contaminants, and hazardous wastes for federal agencies in the United States and Canada, international bodies such as the World Health Organization, and the Ontario Ministry of the Environment. He also has a long record of distinguished service as an officer of Canadian, North American, and international non-governmental organizations in the field of toxicology. In recognition of his scientific achievements he has received many awards and medals from national and international professional organizations.

Dr Plaa has published over 200 reviews and original research articles and has served on the editorial boards of numerous national and international journals. His current editorial board appointments include *Toxicology Letters* and *Food and Chemical Toxicology*.

GITA SEN Born at Poona, India, Dr Sen received a BA in Economics from Fergusson College, University of Poona, in 1968, an MA from the Delhi School of Economics, University of Delhi, in 1970, and a PH D in Economics from Stanford University in 1976. She has held Assistant Professorships in the Department of Economics at Scarborough College, University of Toronto, in 1975/76 and at the New School for Social Research, New York, from 1976 to 1981. In 1981 Dr Sen accepted a position as Associate Fellow, Centre for Development Studies, Trivandrum, India and was appointed a Fellow (Professor) in 1984. She was a Visiting Professor at Harvard University during

1991–93, and is currently a Professor of Economics at the Indian Institute of Management, Bangalore.

Dr Sen has published widely on gender inequalities and economic development in India, and has been an invited participant in numerous international symposia. She received the Volvo Environment Prize in 1994.

A founding member of DAWN (Development Alternatives with Women for a New Era) Dr Sen has served as a consultant to the United Nations and international research projects, and on the editorial boards of *Feminist Studies* and *The Women and Development Annual*.

ALASTAIR M. TAYLOR Born at Vancouver, British Columbia, Dr Taylor obtained his early education in Vancouver and in Hollywood and received BA and MA degrees from the University of Southern California, where he continued with an appointment as lecturer until 1941. Doctoral studies at Columbia University were curtailed in 1942 with his employment at the National Film Board of Canada preparing war training and documentary films. In 1944 he joined the secretariat of the United Nations Relief and Rehabilitation Administration (UNRRA) in Washington, DC. In 1946 he transferred to the Secretariat of the United Nations in New York as a senior editor in the Department of Public Information, a position he held until 1952. In 1949/50, during the transfer of sovereignty from the Netherlands to the Republic of Indonesia, he was Official Spokesman of the Security Council's United Nations Commission for Indonesia. In 1955 he received a D PHIL in International Relations from Oxford University. A Visiting Professorship in the Department of Geography at Edinburgh University in 1959/60 was followed by a joint appointment as Associate Professor in the Departments of Political Studies and Geography at Queen's University at Kingston and promotion to Professor in 1965. Dr Taylor has held visiting Professorships at the University of the West Indies, Bermuda College, and at the University of Guelph, and has taken a sabbatical study leave at the University of Edinburgh. He is the author or co-author of numerous books on history and international politics, including *Civilization: Past and Present* with T.W. Wallbank (1942), now in its eighth edition. Dr Taylor was named Professor Emeritus by Queen's University at Kingston in 1980. Since that time he has continued his academic pursuits and is a consultant to various organizations, including the International Development Research Centre, Ottawa, and the Institute of Noetic Sciences, Sausalito, California. In 1983 he was appointed Adjunct Professor, Antioch-Seattle University, a position he still holds. His long time interests in the environment and in societal history and development continue with his current project, a book analyzing environmental issues in Canada and the feasibility, in this context, of the sustainable development concept put forward in the Brundtland Report.

DUNCAN M. TAYLOR Born in New York, NY, Dr Taylor obtained his early education in Kingston, Ontario, and received a BA degree from Queen's University in 1974. A PH D in 1985 from the University of California at Santa Cruz was followed by an appointment with the Environmental Studies Program at the University of Victoria, where he designed and initiated the core course in the program. Appointed an Assistant Professor in 1989, Dr Taylor helped to initiate the Centre for Sustainable Regional Development at the University of Victoria and is currently collaborating in an ongoing study of the political and environmental options for forestry communities on Vancouver Island. In 1991 he was appointed to the board of the Whistler Foundation for a Sustainable Environment. He is a member of the Government of British Columbia's Old Growth Forests Forum and an adjunct faculty member with Antioch University in Seattle, Washington.

Index